Gröbner Bases in Commutative Algebra

Viviana Ene
Jürgen Herzog

Graduate Studies
in Mathematics
Volume 130

American Mathematical Society
Providence, Rhode Island

2010 *Mathematics Subject Classification.* Primary 13-01, 13A15, 13D02, 13H10, 13P10.

For additional information and updates on this book, visit
www.ams.org/bookpages/gsm-130

Library of Congress Cataloging-in-Publication Data

Ene, Viviana, 1960–
Gröbner bases in commutative algebra / Viviana Ene, Jürgen Herzog.
p. cm. – (Graduate studies in mathematics ; v. 130)
Includes bibliographical references and index.
ISBN 978-0-8218-7287-1 (alk. paper)
1. Gröbner bases. 2. Commutative algebra. I. Herzog, Jürgen, 1947– II. Title.

QA251.3.E54 2012
512′.44–dc23 2011032432

♾ The paper used in this book is acid-free and falls within the guidelines
established to ensure permanence and durability.
Visit the AMS home page at http://www.ams.org/

10 9 8 7 6 5 4 3 2 1 17 16 15 14 13 12

To our parents Maria and Ion, Margarete and Walter

Contents

Preface

Gröbner basis theory has become a fundamental field in algebra which provides a wide range of theoretical and computational methods in many areas of mathematics and other sciences. Bruno Buchberger defined the notion of Gröbner basis in 1965 [**Bu65**]. An intensive research in this theory, related algorithms and applications developed, and many books on this topic have appeared since then. Among them, the books of Adams and Loustaunau [**AL94**], Becker, Kredel, Weispfenning [**BKW93**], Cox, Little, O'Shea [**CLO05**], [**CLO07**], and Eisenbud [**E95**] give a fine introduction to Gröbner basis theory and its applications. Many computer algebra systems like CoCoA, Macaulay2, Magma, Maple, Mathematica, or Singular have implemented various versions of Buchberger's algorithm.

This book aims to provide a concise but rather comprehensive introduction to the theory of Gröbner bases and to introduce the reader to different current trends in theoretical applications of this theory. The complexity level of the presentation increases gradually. The first three chapters and part of Chapter 4 are self-contained and lead to a quick insight into the basics of Gröbner basis theory. They require only a very basic knowledge of algebra. Therefore, this first part of the book would also be appropriate for those readers who are only familiar with elementary algebraic concepts. The second part of Chapter 4 and the last two chapters discuss more advanced topics related to the theory together with applications of it and require a reasonable knowledge in commutative and homological algebra.

Our purpose in writing this book was to provide young researchers and graduate students in commutative algebra and algebraic geometry with methods and techniques related to the Gröbner basis theory. Although it was not our goal to illustrate the algorithmic and computational attributes of the Gröbner basis theory, the usage of the computer in testing examples is indispensable. Users of computer algebra systems may consult specialized monographs like [**GP02**], [**Mac01**] or [**KR00**], [**KR05**].

We give now a brief summary of the book's content. Chapter 1 presents polynomial rings in finitely many indeterminates over a field together with their basic properties and studies ideals in this class of rings. The last two sections are devoted to monomial ideals and standard operations on them. Chapter 2 provides a short but comprehensive exposition of the Gröbner basis notion and Buchberger's criterion and algorithm. In Chapter 3 we discuss first applications based on the Elimination Theorem. Chapter 4 is devoted to the extension of the Gröbner basis theory to submodules of free modules over polynomial rings. The chapter begins with a quick introduction to module theory. The more general concepts discussed here lead to a proof, due to Schreyer, for the celebrated Hilbert's Syzygy Theorem. Chapter 5 opens the series of applications of Gröbner basis theory in commutative algebra and combinatorics. In this chapter we discuss semigroup rings and toric ideals which are intensively studied nowadays from different points of view. Chapter 6 intends to introduce the reader to more advanced topics and to subjects of recent research. The topics treated in this section are not presented in the largest possible generality. Instead, one of the main goals of this last chapter of this monograph is to inspire and to enable the reader, who is interested in further developments and other aspects of the theory, to read more advanced monographs and articles on these subjects, for example, the monograph of Sturmfels [**St95**] which influenced the writing of this chapter substantially, the book [**MS05**] of Miller and Sturmfels, the influential monograph [**S96**] of Stanley and the book of Hibi [**Hi92**]. A compact and detailed presentation of determinantal ideals can be found in the article of Bruns and Conca [**BC03**]. A reader who is interested in further results on monomial ideals should consult the book [**V01**] of Villarreal and the book [**HH10**] of Herzog and Hibi. Further references to research articles are given in the text of Chapter 6. In the first section of the chapter Gröbner bases are used to study Koszul algebras. In particular, it is shown that algebras whose defining ideal has a quadratic Gröbner basis are Koszul. Large classes of algebras whose defining ideal has a quadratic Gröbner basis are provided by algebras generated by sortable sets. This is the topic of the next section. The theory of sortable sets is then applied in the following sections to study generalized Hibi rings and Rees algebras. Next we outline the approach to the theory of determinantal ideals via Gröbner basis with the conclusion that these ideals are all Cohen–Macaulay. Then we give a short introduction to Sagbi bases and apply the theory to show that the coordinate ring of the Grassmannian of m-dimensional vector K-subspaces of K^n is a Gorenstein ring and compute its dimension. The last two sections of the chapter deal with binomial edge ideals and some aspects of algebraic statistics.

The book contains over one hundred problems with a moderate level of difficulty which illuminate the theory and help the reader to fully understand the results discussed in the text. Some of them complete the proofs. Other problems are more complex and encourage the reader to re-examine the simple but essential ideas, to establish connections, and to become interested in further reading.

We wish to express our thanks to Marius Vlădoiu for the careful reading of earlier drafts of this book.

Viviana Ene and Jürgen Herzog

Chapter 1

Polynomial rings and ideals

Gröbner basis theory allows explicit calculations in polynomial rings. In this chapter we introduce polynomial rings in several variables defined over a field K, prove some of their basic properties and study the ideals in these rings. In this book all rings under consideration are commutative and have a unit element.

1.1. Polynomial rings

1.1.1. Definition of the polynomial ring. Let K be a field. We define the polynomial ring $S = K[x_1, \ldots, x_n]$ in n variables over K. The underlying set of this ring has the structure of a K-vector space. The basis elements of this K-vector space are expressions of the form

$$x_1^{a_1} x_2^{a_2} \cdots x_n^{a_n} \quad \text{with} \quad a_i \in \mathbb{N},$$

where we denote by $\mathbb{N}$ the set of nonnegative integers. These expressions are called **monomials**. Thus an arbitrary element in S is a finite linear combination of monomials with coefficients in K. The elements in S are called **polynomials**.

It is customary to omit the factors in a monomial whose exponents are 0, and to view 1 as the monomial with all $a_i = 0$. For example, with this convention, $x_1^3 x_2^0 x_3^2 \in K[x_1, x_2, x_3]$ is written as $x_1^3 x_3^2$. Each monomial is of course also a polynomial.

Here is an example of a polynomial:

$$f = \frac{1}{7}x_1x_2^3x_3^2 - 5x_1^3x_3^5 + \frac{12}{25} \tag{1.1}$$

in the polynomial ring $\mathbb{Q}[x_1, x_2, x_3]$.

We write $\mathbf{x}^{\mathbf{a}}$ for the monomial with exponent vector $\mathbf{a} = (a_1, \ldots, a_n)$. Then a polynomial $f \in S$ can be written as $f = \sum_{\mathbf{a}\in\mathbb{N}^n} c_{\mathbf{a}}\mathbf{x}^{\mathbf{a}}$ with $c_{\mathbf{a}} \in K$ and all but finitely many $c_{\mathbf{a}} = 0$. We set $\operatorname{supp}(f) = \{\mathbf{x}^{\mathbf{a}} \colon\ c_{\mathbf{a}} \neq 0\}$, and call this (finite) set of monomials the **support** of f. Observe that $\operatorname{supp}(f) = \emptyset$ if and only if $f = 0$.

In order to give S a ring structure we have to explain how to add and to multiply polynomials. Let $f, g \in S$ be two polynomials. Since S has the structure of a K-vector space the sum $f + g$ is already defined. Indeed, if $f = \sum_{\mathbf{a}\in\mathbb{N}^n} c_{\mathbf{a}}\mathbf{x}^{\mathbf{a}}$ and $g = \sum_{\mathbf{a}\in\mathbb{N}^n} d_{\mathbf{a}}\mathbf{x}^{\mathbf{a}}$, then

$$f + g = \sum_{\mathbf{a}\in\mathbb{N}^n} (c_{\mathbf{a}} + d_{\mathbf{a}})\mathbf{x}^{\mathbf{a}}.$$

For example, if $g = 5x_1^3x_3^5 + x_1x_2 + 1$ and f is the polynomial in Example (1.1), then $f + g = \frac{1}{7}x_1x_2^3x_3^2 + x_1x_2 + \frac{37}{25}$.

A polynomial of the form $c\mathbf{x}^{\mathbf{a}}$ with $c \in K$ is called a **term**. We first define the multiplication of terms by setting

$$c\mathbf{x}^{\mathbf{a}} \cdot d\mathbf{x}^{\mathbf{b}} = cd\mathbf{x}^{\mathbf{a}+\mathbf{b}}.$$

Since any polynomial is a finite sum of terms and since the multiplication should satisfy the distributive law, the multiplication of any two polynomials is determined. For example, let $f = 2x_1^2 + x_2x_3$ and $g = x_1^2x_3 - 3x_2x_3^2$ be polynomials in $\mathbb{Q}[x_1, x_2, x_3]$. Then

$$\begin{aligned} fg &= (2x_1^2 + x_2x_3)(x_1^2x_3 - 3x_2x_3^2) \\ &= 2x_1^4x_3 - 6x_1^2x_2x_3^2 + x_1^2x_2x_3^2 - 3x_2^2x_3^3 \\ &= 2x_1^4x_3 - 5x_1^2x_2x_3^2 - 3x_2^2x_3^3. \end{aligned}$$

In general, if $f = \sum_{\mathbf{a}\in\mathbb{N}^n} c_{\mathbf{a}}\mathbf{x}^{\mathbf{a}}$ and $g = \sum_{\mathbf{a}\in\mathbb{N}^n} d_{\mathbf{a}}\mathbf{x}^{\mathbf{a}}$, then

$$fg = \sum_{\mathbf{g}\in\mathbb{N}^n} \Big(\sum_{\mathbf{a}+\mathbf{b}=\mathbf{g}} c_{\mathbf{a}}d_{\mathbf{b}} \Big)\mathbf{x}^{\mathbf{g}}.$$

The reader easily checks that S with the addition and multiplication so defined is indeed a ring.

A K**-algebra** is a ring containing the field K as a subring. Our polynomial ring S is a K-algebra, since we may identify the constant polynomials in S with the field K. The nonzero constant polynomials are those whose support is the set $\{1\}$.

Let A and B be K-algebras. A ring homomorphism $\varphi\colon A \to B$ is called a **K-algebra homomorphism**, if $\varphi(a) = a$ for all $a \in K$. In other words, the restriction $\varphi|_K$ of φ to K is the identity map.

The polynomial ring has the following nice universal mapping property.

Theorem 1.1. *Let R be a K-algebra, and let $\alpha_1, \dots, \alpha_n$ be arbitrary elements in R. Then there is a unique K-algebra homomorphism*

$$\varphi\colon S \longrightarrow R$$

with the property that $\varphi(x_i) = \alpha_i$ for $i = 1, \dots, n$.

Proof. Suppose there exists a K-algebra homomorphism $\varphi\colon S \to R$ with $\varphi(x_i) = \alpha_i$ for $i = 1, \dots, n$, and let $f = \sum_{\mathbf{a}\in\mathbb{N}} c_{\mathbf{a}}\mathbf{x}^{\mathbf{a}}$ be an arbitrary polynomial in S. Then, since φ is a K-algebra homomorphism, it follows that

$$\varphi(f) = \sum_{\mathbf{a}\in\mathbb{N}^n} \varphi(c_{\mathbf{a}})\varphi(\mathbf{x}^{\mathbf{a}}) = \sum_{\mathbf{a}\in\mathbb{N}^n} c_{\mathbf{a}}\alpha^{\mathbf{a}}. \tag{1.2}$$

This calculation shows that if such a K-algebra homomorphism exists, then it is uniquely determined. Hence we have no other choice but to define φ as in (1.2). The map φ is therefore defined by substituting the variables x_i in the polynomials by the α_i. This substitution is compatible with the ring operations and is the identity on the constant polynomials. Therefore, φ is a K-algebra homomorphism. □

The K-algebra homomorphism φ defined in the proof of Theorem 1.1 is called the **substitution homomorphism**.

The set of monomials in $K[x_1]$ is $\{x_1^0 = 1, x_1, x_1^2, \dots\}$. This set of monomials is naturally ordered according to the exponents of x_1. For $n > 1$ there is a priori no order of the monomials given. But at least we can define a natural partial order given by the total degree of the monomials which is defined as follows. Let $\mathbf{x}^{\mathbf{a}} = x_1^{a_1}\cdots x_n^{a_n}$ be a monomial. Then we set

$$\deg \mathbf{x}^{\mathbf{a}} = |\mathbf{a}|, \quad \text{where} \quad |\mathbf{a}| = \sum_{i=1}^n a_i,$$

and define the **degree** of an arbitrary polynomial to be the number

$$\deg f = \max\{\deg \mathbf{x}^{\mathbf{a}}\colon\ \mathbf{x}^{\mathbf{a}} \in \operatorname{supp}(f)\}.$$

The monomials in the support of the polynomial f in Example (1.1) have different degrees. The highest degree among them is 8, and hence $\deg f = 8$ in this example.

A polynomial f is said to be **homogeneous** of degree i, if all monomials in the support of f are of degree i. Let $f = \sum_{\mathbf{a}\in\mathbb{N}^n} c_{\mathbf{a}}\mathbf{x}^{\mathbf{a}}$ be an arbitrary polynomial. We set

$$f_i = \sum_{\mathbf{a}\in\mathbb{N}^n,\ |\mathbf{a}|=i} c_{\mathbf{a}}\mathbf{x}^{\mathbf{a}}.$$

Then f_i is homogeneous of degree i and is called the **i-th homogeneous component** of f. One has $f = \sum_{i\geq 0} f_i$, and this decomposition into homogeneous components is unique. It follows that

$$S = \bigoplus_{j=0}^{\infty} S_j,$$

where S_j is the K-subspace of S consisting of all homogeneous polynomials in S of degree j.

Let R be an arbitrary commutative ring. A nonempty subset $I \subset R$ is called an **ideal**, if $f+g \in I$ and $hf \in I$ for all $f, g \in I$ and $h \in R$.

Ideals naturally occur as kernels of ring homomorphisms. Let $\varphi\colon R \to R'$ be a ring homomorphism. The **kernel** of φ is defined to be the ideal

$$\operatorname{Ker}(\varphi) = \{r \in R\colon\ \varphi(r) = 0\}.$$

It is easily verified that $\operatorname{Ker}(\varphi)$ is indeed an ideal in R.

In geometric context, ideals appear as sets of all the polynomials vanishing on a given subset of K^n, as we will explain later, in Section 3.3.

Given an arbitrary subset $\mathcal{G} \subset R$, we define $I = (\mathcal{G})$ to be the set of all linear combinations of elements in $\mathcal{G}$, that is, the set of all finite sums $\sum_{i=1}^{s} g_i f_i$ with $f_i \in \mathcal{G}$ and $g_i \in R$. This set I is indeed an ideal, and is called the ideal generated by the set $\mathcal{G}$. If $\mathcal{G}$ is a finite set, say, $\mathcal{G} = \{f_1, \ldots, f_m\}$, then we write $I = (f_1, \ldots, f_m)$ and say that I is **generated** by $f_1, \ldots, f_m$.

An ideal I in the polynomial ring $S = K[x_1, \ldots, x_n]$ is said to be **graded**, if I is generated by homogeneous polynomials.

Proposition 1.2. *Let $I \subset S$ be an ideal. The following conditions are equivalent:*

(a) *I is a graded ideal;*

(b) *if $f \in I$, then all homogeneous components f_j of f belong to I;*

(c) *$I = \bigoplus_{j=0}^{\infty} I_j$ where $I_j = I \cap S_j$.*

Proof. (a) $\Rightarrow$ (b): Let $\mathcal{G}$ be a homogeneous system of generators of I, and let $f \in I$ be an arbitrary polynomial. Then there exists $g_1, \ldots, g_m$ in $\mathcal{G}$ and $h_1, \ldots, h_m \in S$ such that $f = \sum_{i=1}^{m} h_i g_i$. Let $a_i = \deg g_i$; then the jth homogeneous component f_j of f is equal to $\sum_{i=1}^{m} k_i g_i$, where k_i is the $(j-a_i)$th component of h_i. In particular, $f_j \in I$.

(b) $\Rightarrow$ (c): Let $f \in I$. Then $f_j \in I$, and hence $f_j \in I \cap S_j$. This shows that $I = \sum_{j=0}^{\infty} I_j$. Since the decomposition of a polynomial into homogeneous components is unique, we see that $\sum_{j=0}^{\infty} I_j = \bigoplus_{j=0}^{\infty} I_j$.

(c) $\Rightarrow$ (b): Obvious.

(b) $\Rightarrow$ (a): Let $\mathcal{G}$ be an arbitrary system of generators of I, and for each $g \in \mathcal{G}$ let g_j be the jth homogeneous component of g. Then $\mathcal{F} = \{g_j : g \in \mathcal{G}, j = 0, 1, \ldots\}$ is a homogeneous system of generators of I. □

1.1.2. Some basic properties of polynomial rings. The polynomial ring $R[x]$ in one variable over a ring R is known from basic courses in algebra. A polynomial $f \in R[x]$ is an expression of the form $f = \sum_{i=0}^{n} r_i x^i$, where $n > 0$ is an integer and $r_i \in R$ for $i = 0, \ldots, n$. The **degree** of a polynomial $f \neq 0$, denoted by $\deg f$, is the number $\max\{i\colon\ r_i \neq 0\}$. If $f = 0$, we set $\deg f = -\infty$.

We now consider the special case that $R = K[x_1, \ldots, x_{n-1}]$ and consider the polynomial ring $R[x_n]$ in one variable over this ring R. Since R is a K-algebra, $R[x_n]$ is a K-algebra as well. As before we set $S = K[x_1, \ldots, x_n]$.

Theorem 1.3. *There is a natural isomorphism $S \cong R[x_n]$ of K-algebras.*

Proof. Let $f \in S$. Then f can be uniquely written as $f = \sum_{i \geq 0} f_i x_n^i$ with $f_i \in R$. Thus we may view f also as an element of $R[x_n]$. The K-algebra isomorphism $S \to R[x_n]$ is then just the expansion of f with respect to the powers of x_n as described above. One has to verify that this assignment is compatible with the K-algebra structures on both sides. □

This simple result has the following nice consequence.

Corollary 1.4. *The polynomial ring $S = K[x_1, \ldots, x_n]$ is a factorial domain.*

Proof. By using Theorem 1.3, we may identify S with $R[x_n]$, where $R = K[x_1, \ldots, x_{n-1}]$. Proceeding by induction on n, we may assume that R is a factorial domain. Thus as a consequence of the lemma of Gauss on primitive polynomials, known from the algebra courses, the assertion follows. □

Another important consequence of Theorem 1.3 is

Corollary 1.5 (Hilbert's basis theorem). *The ring $S = K[x_1, \ldots, x_n]$ is Noetherian. In other words, each ideal in S is finitely generated.*

The more general version of Hilbert's basis theorem says that the polynomial ring $R[x]$ over R is Noetherian, if R is Noetherian. Corollary 1.5 follows from this fact and Theorem 1.3 by using induction on the number of the variables. A different proof of Corollary 1.5 will be given later in Section 2.2 by using Gröbner basis arguments.

1.2. Ideals

1.2.1. Operations on ideals. In the previous section we have already introduced ideals. Here we discuss some algebraic operations on ideals. Let R be an arbitrary commutative ring, and let I and J be ideals in R. The **sum** and **product** of the two ideals is defined as follows:

$$I + J = \{f + g : f \in I, g \in J\},$$

and

$$IJ = (\mathcal{G}), \quad \text{where} \quad \mathcal{G} = \{fg : f \in I, g \in J\}.$$

The product IJ is by its definition an ideal. But obviously the sum $I + J$ is an ideal, too. Let I, J, K be ideals. Then:

(i) $IJ = JI$ and $I + J = J + I$;

(ii) $(IJ)K = I(JK)$ and $(I + J) + K = I + (J + K)$;

(iii) $I(J + K) = IJ + IK$.

Observe that the intersection $I \cap J$ of the ideals I and J is again an ideal, and the following rules hold:

(iv) $IJ \subseteq I \cap J$;

(v) $I \cap J + I \cap K \subseteq I \cap (J + K)$.

The **ideal quotient** of the ideals I and J is defined to be the ideal

$$I : J = \{f \in R : fg \in I \text{ for all } g \in J\}.$$

The following rules hold:

(vi) $I : JK = (I : J) : K$;

(vii) $I : (J + K) = (I : J) \cap (I : K)$;

(viii) $I : K + J : K \subseteq (I + J) : K$.

An ideal $P \subset R, P \neq R$, is called **prime** if it satisfies the following condition: for any $x, y \in R$, if $xy \in P$, then $x \in P$ or $y \in P$. An ideal $M \subset R, M \neq R$, is called **maximal** if it is a maximal element (with respect to inclusion) of the set of all ideals properly contained in R. By Problem 1.14, the ring $S = K[x_1, \ldots, x_n]$ has a unique graded maximal ideal, namely $(x_1, \ldots, x_n)$.

Let $I \subset S = K[x_1, \ldots, x_n]$ be a graded ideal and $\mathfrak{m} = (x_1, \ldots, x_n)$ the graded maximal ideal of S. The ideal I is called **saturated**, if $I : \mathfrak{m} = I$.

We introduce the ideal I^{sat} to be $I : \mathfrak{m}^\infty = \bigcup_{i=1}^{\infty} I : \mathfrak{m}^i$, and call it the **saturation** of I. This naming is justified by the next result.

Proposition 1.6. *Let $I, J \subset S$ be graded ideals. Then the following hold:*

(i) I^{sat} *is saturated;*

(ii) *if* $I \subset J$, *then* $I^{\text{sat}} \subset J^{\text{sat}}$*;*

(iii) *if* I *and* J *are saturated, then* $I \cap J$ *is saturated;*

Proof. (i) Suppose that $f\mathfrak{m} \in I : \mathfrak{m}^\infty$. Then there exists an integer k such that $f\mathfrak{m} \in I : \mathfrak{m}^k$. In other words, $f \in (I : \mathfrak{m}^k) : \mathfrak{m} = I : \mathfrak{m}^{k+1} \subseteq I : \mathfrak{m}^\infty$.
(ii) Is obvious.
(iii) Suppose that $f\mathfrak{m}^k \subseteq I \cap J$ for some k. Then $f\mathfrak{m}^k \subseteq I$ and $f\mathfrak{m}^k \subseteq J$. Since I and J are saturated, it follows that $f \in I$ and $f \in J$, and hence $f \in I \cap J$. $\square$

Finally, we consider the **radical** $\sqrt{I}$ of an ideal I. We define

$$\sqrt{I} = \{f \in R : \ f^k \in I \quad \text{for some integer} \quad k > 0\}.$$

$\sqrt{I}$ is indeed an ideal. To see this, let $f, g \in \sqrt{I}$ and $h \in R$. Then $f^k \in I$ and $g^l \in I$ for some integers $k, l > 0$. Therefore, $(hf)^k = h^k f^k \in I$. This shows that $hf \in \sqrt{I}$. Furthermore, $(f+g)^{k+l} = \sum_{i=0}^{k+l} \binom{k+l}{i} f^i g^{k+l-i} \in I$, because for each $i = 0, \dots, k+l$ either $f^i \in I$ or $g^{k+l-i} \in I$. This shows that $f + g \in \sqrt{I}$.

An ideal I is said to be a **radical ideal** if $I = \sqrt{I}$. Since $\sqrt{\sqrt{I}} = \sqrt{I}$, it follows that $\sqrt{I}$ is a radical ideal. In fact, it is the smallest radical ideal which contains I.

1.2.2. Residue class rings. Let $I \subset R$ be an ideal, and f an element in R. The set $f + I = \{f + h \colon h \in I\}$ is called the **residue class of f modulo** I, and f is called a **representative** of the residue class $f + I$. Observe that $f + I = g + I$, if and only if $f - g \in I$. Thus different representatives of a residue class differ only by an element of I.

The set of residue classes modulo I is denoted by R/I. We give R/I a ring structure by defining the addition and multiplication on R/I as follows:

$$(f + I) + (g + I) = (f + g) + I \quad \text{and} \quad (f + I)(g + I) = (fg) + I.$$

The reader checks easily that the definitions do not depend on the particular chosen representatives and that R/I with the so-defined addition and multiplication is indeed a ring. The natural map $\epsilon \colon R \to R/I$ with $\epsilon(f) = f + I$ is a surjective ring homomorphism, called the **canonical epimorphism**, and one has $\text{Ker}(\epsilon) = I$.

It is well known from the basic algebra courses that the following equivalences hold.
(i) $P \subset R$ is a prime ideal if and only if R/P is a domain.
(ii) $M \subset R$ is a maximal ideal if and only if R/M is a field.

1.2.3. Monomial ideals and Dickson's lemma. Let $S = K[x_1, \ldots, x_n]$ be the polynomial ring over a field K. A **monomial ideal** in S is an ideal generated by monomials. This class of ideals is of interest to us, because Gröbner basis theory reduces difficult algebraic calculations to calculations with monomial ideals, which are essentially of combinatorial nature.

Given two monomials $u = x_1^{a_1} \cdots x_n^{a_n}$ and $v = x_1^{b_1} \cdots x_n^{b_n}$ in S, then u **divides** v, if $a_i \leq b_i$ for all i. In this case we write $u|v$. Since S is factorial, for any two polynomials $f, g \in S$, there exists the greatest common divisor $\gcd(f, g)$ and the least common multiple $\operatorname{lcm}(f, g)$. We have

$$\gcd(u, v) = x_1^{\min\{a_1,b_1\}} \cdots x_n^{\min\{a_n,b_n\}}$$

and

$$\operatorname{lcm}(u, v) = x_1^{\max\{a_1,b_1\}} \cdots x_n^{\max\{a_n,b_n\}}.$$

The following characterization of monomial ideals is of fundamental importance.

Theorem 1.7. *Let $I \subset S$ be an ideal. The following conditions are equivalent:*

(a) *I is a monomial ideal;*

(b) *for any $f \in I$, one has $\operatorname{supp}(f) \subseteq I$.*

Proof. (a) $\Rightarrow$ (b): Let $\mathcal{M}$ be a set of monomial generators of I, and let $f \in I$. Then there exist $u_1, \ldots, u_m \in \mathcal{M}$ and $f_1, \ldots, f_m \in S$ such that $f = \sum_{i=1}^m f_i u_i$. It follows that $\operatorname{supp}(f) \subseteq \bigcup_{i=1}^m \operatorname{supp}(f_i u_i)$. Thus if $u \in \operatorname{supp}(f)$, then there exists i such that $u \in \operatorname{supp}(f_i u_i)$. Since each $v \in \operatorname{supp}(f_i u_i)$ is of the form wu_i, it follows that $u = wu_i$ for some i and some monomial w. Hence $u \in I$.

(b) $\Rightarrow$ (a): Let $\mathcal{G}$ be any system of generators of I. Then the set $\bigcup_{f \in \mathcal{G}} \operatorname{supp}(f)$ is contained in I and is a set of monomial generators of I. □

Corollary 1.8. *Let $I \subset S$ be a monomial ideal, and $\mathcal{M}$ a set of monomials in I. Then $\mathcal{M}$ is a set of generators of I, if and only if for each monomial $v \in I$ there exists $u \in \mathcal{M}$ such that $u|v$.*

Proof. Assume $\mathcal{M}$ is a set of generators of I. Let $v \in I$ be a monomial. There exist $u_1, \ldots, u_m \in \mathcal{M}$ and $f_1, \ldots, f_m \in S$ such that $v = \sum_{i=1}^m f_i u_i$. Therefore $v \in \bigcup_{i=1}^m \operatorname{supp}(f_i u_i)$, and hence $v \in \operatorname{supp}(f_i u_i)$ for some i. This implies that u_i divides v.

Conversely, suppose that for each monomial $v \in I$ there exists $u \in \mathcal{M}$ such that $u|v$. Let $f \in I$ be an arbitrary polynomial. Since I is a monomial ideal, Theorem 1.7 implies that $\operatorname{supp}(f) \subset I$. Let $\operatorname{supp}(f) = \{v_1, \ldots, v_m\}$ and $f = \sum_{i=1}^m c_i v_i$ with $c_i \in K$. By assumption, $v_i = w_i u_i$ with $u_i \in \mathcal{M}$ and

w_i a monomial in S. It follows that $f = \sum_{i=1}^m c_i w_i u_i$. This shows that $\mathcal{M}$ generates I. □

Next we will show that each monomial ideal has a finite set of monomial generators. To this end we first need to prove

Theorem 1.9 (Dickson's lemma). *Let $\mathcal{M}$ be a nonempty set of monomials in S. Then with respect to the partial order given by divisibility, the set $\mathcal{M}$ has only a finite number of minimal elements.*

Proof. We prove Dickson's lemma by induction on n, the number of variables of S. If $n = 1$, then $\mathcal{M}$ consists of certain powers of x_1, and the set of minimal elements of $\mathcal{M}$ is the set $\{x_1^a\}$, where a is the smallest number such that $x_1^a \in \mathcal{M}$.

Now let $n > 1$, and let $\mathcal{N}$ be the set of monomials $\mathbf{x^c} \in K[x_1, \ldots, x_{n-1}]$ such that $\mathbf{x^c} x_n^d \in \mathcal{M}$ for some $d \geq 0$. By induction hypothesis, the set $\mathcal{N}^{\min}$ of minimal elements of $\mathcal{N}$ is finite, say $\mathcal{N}^{\min} = \{\mathbf{x}^{\mathbf{c}_1}, \ldots, \mathbf{x}^{\mathbf{c}_r}\}$. For each $\mathbf{x}^{\mathbf{c}_i}$ there exists $a_i \geq 0$ such that $\mathbf{x}^{\mathbf{c}_i} x_n^{a_i} \in \mathcal{M}$. Let $a = \max\{a_1, \ldots, a_r\}$, and for each b with $0 \leq b < a$ let $\mathcal{N}_b = \{\mathbf{x^c} \in K[x_1, \ldots, x_{n-1}] : \mathbf{x^c} x_n^b \in \mathcal{M}\}$. Again by induction hypothesis, $\mathcal{N}_b^{\min}$ is a finite set. We denote the set of monomials $\mathbf{x^c} x_n^b$ with $\mathbf{x^c} \in \mathcal{N}_b^{\min}$ by $\mathcal{N}_b^{\min} x_n^b$, and claim that

$$\mathcal{M}^{\min} \subseteq \{\mathbf{x}^{\mathbf{c}_1} x_n^{a_1}, \ldots, \mathbf{x}^{\mathbf{c}_r} x_n^{a_r}\} \cup \bigcup_{b=0}^{a-1} \mathcal{N}_b^{\min} x_n^b.$$

Since the right-hand side of this inclusion is a finite set, the assertion of the theorem follows from this claim.

In order to prove the claim, let $u = \mathbf{x^c} x_n^d$ be a monomial in $\mathcal{M}$. If $d \geq a$, then some monomial in $\{\mathbf{x}^{\mathbf{c}_1} x_n^{a_1}, \ldots, \mathbf{x}^{\mathbf{c}_r} x_n^{a_r}\}$ divides u. If $0 \leq d < a$, then u is divisible by a monomial in $\mathcal{N}_d^{\min} x_n^d$, as desired. □

Now we give a direct proof of the Hilbert's basis theorem (Corollary 1.5) for monomial ideals.

Corollary 1.10. *Let I be a monomial ideal. Then each set of monomial generators of I contains a finite set which generates I.*

Proof. Let $\mathcal{M}$ be a set of monomial generators of I. By Dickson's lemma, the set of minimal elements of $\mathcal{M}$ is finite. This finite set is a set of monomial generators of I. □

Let I be a monomial ideal. A set of monomial generators of I is called **minimal** if any proper subset of it is *not* a set of generators of I. By the preceding corollary, each minimal set of monomial generators of I is finite. Moreover, we have

Proposition 1.11. *Let I be a monomial ideal. Then there exists a unique minimal set of monomial generators of I.*

Proof. The existence of a minimal set of monomial generators of I follows from Corollary 1.10. Now let $\{u_1, \ldots, u_r\}$ and $\{v_1, \ldots, v_s\}$ be minimal monomial set of generators of I. Since $I = (v_1, \ldots, v_s)$ and since $u_i \in I$, it follows from Corollary 1.8 that $v_j | u_i$ for some j. Similarly, $u_k | v_j$ for some k, and hence $u_k | u_i$. Since $\{u_1, \ldots, u_r\}$ is a minimal monomial set of generators of I, we see that $i = k$, and hence $u_i = v_j$. This implies $\{u_1, \ldots, u_r\} \subseteq \{v_1, \ldots, v_s\}$. Similarly, $\{v_1, \ldots, v_s\} \subseteq \{u_1, \ldots, u_r\}$. □

Henceforth, the unique minimal set of monomial generators of I will be denoted by $G(I)$.

As another consequence of Dickson's lemma we have

Proposition 1.12. *Each ascending sequence of monomial ideals*

$$I_1 \subseteq I_2 \subseteq \cdots \subseteq I_j \subseteq \cdots$$

in S terminates, that is, there exists an integer k such that $I_\ell = I_k$ for all $\ell \geq k$.

Proof. Let $\mathcal{M} = \bigcup_{j=1}^{\infty} G(I_j)$. According to Dickson's lemma, the set $\mathcal{M}^{\min}$ is finite. Hence there is an integer k such that $\mathcal{M}^{\min} \subseteq \bigcup_{j=1}^{k} G(I_j)$. Now let $\ell \geq k$ and let u be a monomial in I_ℓ. Then there exists $v \in \mathcal{M}^{\min}$ which divides u. This implies that $u \in \bigcup_{j=1}^{k} I_j = I_k$, as desired. □

1.2.4. Operations on monomial ideals. Let I and J be monomial ideals. Then $I + J$ and IJ are again monomial ideals with $G(I+J) \subseteq G(I) \cup G(J)$ and $G(IJ) \subseteq G(I)G(J)$, where $G(I)G(J) = \{uv \,:\, u \in G(I),\ v \in G(J)\}$.

For example, let $I = (x_1x_2, x_2x_3^2)$ and $J = (x_1^2x_2, x_2x_3)$. Then by Problem 1.5,

$$I + J = (x_1x_2, x_2x_3^2, x_1^2x_2, x_2x_3) \quad \text{and} \quad G(I+J) = \{x_1x_2, x_2x_3\},$$

and

$$IJ = (x_1^3x_2^2, x_1x_2^2x_3, x_1^2x_2^2x_3^2, x_2^2x_3^3) \quad \text{and} \quad G(IJ) = \{x_1^3x_2^2, x_1x_2^2x_3, x_2^2x_3^3\}.$$

The intersection of two monomial ideals can be computed as follows.

Proposition 1.13. *Let I and J be monomial ideals with $G(I) = \{u_1, \ldots, u_r\}$ and $G(J) = \{v_1, \ldots, v_s\}$. Then*

$$I \cap J = (\{\mathrm{lcm}(u_i, v_j) \,:\, i = 1, \ldots, r,\ j = 1, \ldots, s\}).$$

Proof. We first observe that $I \cap J$ is a monomial ideal. Indeed, let $f \in I \cap J$. Then Theorem 1.7 implies that $\text{supp}(f) \in I$ and $\text{supp}(f) \in J$, since both I and J are monomial ideals. It follows that $\text{supp}(f) \in I \cap J$, and so, again applying Theorem 1.7, we see that $I \cap J$ is a monomial ideal.

Now let $u \in I \cap J$ be a monomial. Then by Corollary 1.8 there exists $u_i \in G(I)$ and $v_j \in G(J)$ such that $u_i | u$ and $v_j | u$. Therefore, $\text{lcm}(u_i, v_j) | u$, and hence $u \in (\{\text{lcm}(u_i, v_j) \ : \ i = 1, \ldots, r, \ j = 1, \ldots, s\})$. On the other hand, since $\text{lcm}(u_i, v_j)$ is a multiple of u_i and a multiple of v_j, it follows that $\text{lcm}(u_i, v_j) \in I \cap J$. $\square$

The ideal quotient of two monomial ideals can also be easily computed.

Proposition 1.14. *Let I and J be monomial ideals with $G(I) = \{u_1, \ldots, u_r\}$ and $G(J) = \{v_1, \ldots, v_s\}$. Then $I : J = \bigcap_{j=1}^{s} I : (v_j)$, and*

$$I : (v_j) = (\{u_i / \gcd(u_i, v_j) \ : \ i = 1, \ldots, r\}).$$

Proof. Similarly, as in the proof of Proposition 1.13, one deduces from Theorem 1.7 that the ideal quotient of two monomial ideals is again a monomial ideal.

The fact that $I : J = \bigcap_{j=1}^{s} I : (v_j)$ follows from Problem 1.7. Now let $u \in I : (v_j)$. Then $uv_j \in I$, and hence there exists i such that $u_i | uv_j$. This implies that $u_i / \gcd(u_i, v_j)$ divides u. In other words, $u \in (\{u_i / \gcd(u_i, v_j) \ : \ i = 1, \ldots, r\})$.

On the other hand, we have $u_i / \gcd(u_i, v_j) \in I : (v_j)$ for all i, since u_i divides $(u_i / \gcd(u_i, v_j))v_j$. $\square$

By using Proposition 1.13 and Proposition 1.14 we are now able to compute the ideal quotient of two monomial ideals. For example, let $I = (x_1x_2^2, x_2x_3^2)$ and $J = (x_1x_2x_3, x_2^2)$. Then

$$\begin{aligned} I : J &= (x_1x_2^2, x_2x_3^2) : (x_1x_2x_3, x_2^2) \\ &= (x_1x_2^2, x_2x_3^2) : (x_1x_2x_3) \cap (x_1x_2^2, x_2x_3^2) : (x_2^2) \\ &= (x_2, x_3) \cap (x_1, x_3^2) = (x_1x_2, x_1x_3, x_3^2). \end{aligned}$$

Finally, we want to discuss how to compute the radical of a monomial ideal. We say that a monomial $\mathbf{x}^{\mathbf{a}}$ is **squarefree** if $a_i \leq 1$ for $i = 1, \ldots, n$. A monomial ideal I is called a **squarefree monomial ideal** if all elements in $G(I)$ are squarefree monomials. According to Problem 1.19, a monomial ideal P is a prime ideal if and only if P is generated by a subset of the variables. Thus monomial prime ideals are a special class of squarefree monomial ideals. Since by Problem 1.20 an intersection of squarefree monomial ideals is again a squarefee monomial ideal, it follows in particular that the intersection of monomial prime ideals is a squarefree monomial ideal. The converse is true as well. Indeed, we have

Proposition 1.15. *Let I be a squarefree monomial ideal. Then I is a finite intersection of monomial prime ideals.*

Proof. Let $G(I) = \{u_1, \ldots, u_r\}$. We prove the assertion by induction on r. If $r = 1$, then $I = (u_1) = (x_{i_1} \cdots x_{i_d}) = \bigcap_{j=1}^d (x_{i_j})$. Now assume $r > 1$ and that $u_1 = x_{i_1} \cdots x_{i_d}$. Then by Proposition 1.13 we have $I = \bigcap_{j=1}^d (x_{i_j}, u_2, \ldots, u_r)$. By induction hypothesis, $(u_2, \ldots, u_r) = \bigcap_{i=1}^s P_i$, where each P_i is a monomial prime ideal. We then get $I = \bigcap_{j=1}^d \bigcap_{i=1}^s ((x_{i_j}) + P_i)$. $\square$

Corollary 1.16. *Let I be a monomial ideal. Then I is a radical ideal if and only if I is a squarefree monomial ideal.*

Proof. Assume that I is a squarefree monomial ideal. Then by Proposition 1.15, I is a finite intersection of monomial prime ideals. Hence by Problem 1.13, I is a radical ideal.

Conversely, assume that I is not squarefree. Then there exists an element $u = x_{i_1}^{a_1} \cdots x_{i_r}^{a_r} \in G(I)$ with $i_1 < \cdots < i_r$, $a_i > 0$ for all i and $a_j > 1$ for some j. Let $a = \max\{a_1, \ldots, a_r\}$. Then $(x_{i_1} \cdots x_{i_r})^a \in I$. Hence $x_{i_1} \cdots x_{i_r} \in \sqrt{I}$. On the other hand, since $u = x_{i_1}^{a_1} \cdots x_{i_r}^{a_r} \in G(I)$, one has $x_{i_1} \cdots x_{i_r} \notin I$. Therefore, I is not a radical ideal. $\square$

Let $u = \mathbf{x}^{\mathbf{a}}$ be a monomial. We set $\sqrt{u} = \prod_{i=1,\ a_i \neq 0}^n x_i$. For example, if $u = x_1 x_3^4 x_4^2$, then $\sqrt{u} = x_1 x_3 x_4$.

Theorem 1.17. *Let I be a monomial ideal. Then*

$$\sqrt{I} = (\sqrt{u} \ : \ u \in G(I)).$$

Proof. Let $J = (\sqrt{u} \ : \ u \in G(I))$. Then $I \subseteq J$ and J is a radical ideal according to Corollary 1.16. Let $f \in \sqrt{I}$. Then $f^m \in I$ for some integer $m > 0$, and so $f^m \in J$. Since J is a radical ideal it follows that $f \in J$. This shows that $\sqrt{I} \subseteq J$.

As in the proof of Corollary 1.16, it follows that for each $u \in G(I)$ there exists an integer $\alpha_u > 0$ with $(\sqrt{u})^{\alpha_u} \in I$. This implies that $J \subseteq \sqrt{I}$. $\square$

Problems

In the following problems, K is a field and S denotes the polynomial ring $K[x_1, \ldots, x_n]$.

Problem 1.1. Let A and B be subsets of S. We set $AB = \{fg\colon f \in A, g \in B\}$. For any two polynomials $f, g \in S$, show that

(i) $\operatorname{supp}(f+g) \subseteq \operatorname{supp}(f) \cup \operatorname{supp}(g)$;

(ii) $\operatorname{supp}(fg) \subseteq \operatorname{supp}(f)\operatorname{supp}(g)$.

Give examples for which these inclusions are strict.

Problem 1.2. Let A be an $n \times n$-matrix with entries a_{ij} in K and let $\alpha_i = \sum_{j=1}^{n} a_{ji}x_j$. Furthermore, let $\varphi\colon\ S \to S$ be the substitution homomorphism with $\varphi(x_i) = \alpha_i$. Show that φ is an isomorphism if and only if A is invertible.

Problem 1.3. Let $f, g \in S$ be two polynomials. Express the i-th homogeneous component of fg in terms of the homogeneous components of f and the homogeneous components of g.

Problem 1.4. Show that the two ideals $(x_1 + x_2, x_2^2)$ and $(x_1 + x_2, x_1^2)$ coincide.

Problem 1.5. Let $I = (f_1, \ldots, f_r)$ and $J = (g_1, \ldots, g_s)$ be two ideals. Show that

$$I + J = (f_1, \ldots, f_r, g_1, \ldots, g_s)$$

and

$$IJ = (f_i g_j\colon\ i = 1, \ldots, r, j = 1, \ldots, s).$$

Problem 1.6. Give an example of ideals I, J, K for which the inclusions $IJ \subseteq I \cap J$ and $I \cap J + I \cap K \subseteq I \cap (J + K)$ are strict.

Problem 1.7. Let I and J be ideals and suppose that $J = (g_1, \ldots, g_m)$. Show that $I : J = \bigcap_{i=1}^{m} I : (g_i)$.

Problem 1.8. Let I be a graded ideal. Show that I^{sat} and $\sqrt{I}$ are again graded ideals.

Problem 1.9. Let $I \subset S$ be a graded ideal. Prove the equality $I^{\text{sat}} = \bigcap_{i=1}^{n} (I\colon x_i^{\infty})$.

Problem 1.10. Let R be an arbitrary ring and I an ideal of R. An element $f \in R$ is called **R/I-regular** if, for any $g \in R$, $f(g+I) = 0$ implies $g+I = 0$, that is, $fg \in I$ implies $g \in I$. $f_1, \ldots f_m \in R$ is an **R/I-sequence** if f_i is $R/(I, f_1, \ldots, f_{i-1})$-regular for all $i \geq 1$. Show that the following conditions are equivalent:

(a) $f_1, \ldots, f_m$ is an R/I-sequence;

(b) $(I, f_1, \ldots, f_{i-1})\colon (f_i) = (I, f_1, \ldots, f_{i-1})$ for all $i \geq 1$.

Problem 1.11. Show that the intersection of radical ideals is a radical ideal.

Problem 1.12. Let $I = (xy, xz, yz)$. Show that I is saturated and that I^2 is not saturated.

Problem 1.13. Let $I = \bigcap_{i=1}^{k} P_i$, where each P_i is a prime ideal. Then I is a radical ideal. (In fact, in a Noetherian ring the converse is also true, but need not be proved here).

Problem 1.14. Prove that the ideal $(x_1, \ldots, x_n)$ is the only graded maximal ideal of the ring $K[x_1, \ldots, x_n]$.

Problem 1.15. Let I be a graded radical ideal different from $\mathfrak{m} = (x_1, \ldots, x_n)$. Prove that I is saturated.

Problem 1.16. Let I be an ideal in $K[x, y]$ generated by $k + 1$ K-linearly independent homogeneous polynomials of degree k. Is I a monomial ideal?

Problem 1.17. Let I, J and K be monomial ideals. Show that

$$I \cap (J + K) = (I \cap J) + (I \cap K).$$

Problem 1.18. Let I be a monomial ideal and let $J = (x_1, \ldots, x_r)$. Show that $I : J \neq I$ if there exist integers $a_i > 0$ such that $x_i^{a_i} \in G(I)$ for $i = 1, \ldots, r$. Is the converse true as well?

Problem 1.19. Show that a monomial ideal P is a prime ideal if and only if P is generated by a subset of the variables.

Problem 1.20. Let I, J be squarefree monomial ideals. Prove that $I \cap J$ is a squarefee monomial ideal.

Chapter 2

Gröbner bases

Given polynomials $f, g \in \mathbb{Q}[x]$ with $g \neq 0$, there exist uniquely determined polynomials q and r in $\mathbb{Q}[x]$ such that $f = qg + r$ and $\deg r < \deg g$. The polynomial r is called the remainder of f with respect to g.

This fundamental fact, known from high school, can be generalized to polynomial rings over an arbitrary field. Indeed, there is an algorithm, known as the Euclidian algorithm, to compute q and r. The algorithm works as follows: if $\deg f < \deg g$, then $q = 0$ and $r = f$. If $\deg f \geq \deg g$, let ax^n with $a \in K$ be the leading term of f and bx^m with $b \in K$ the leading term of g. Then the degree of $r_1 = f - (a/b)x^{n-m}g$ is less than $\deg f$. If $\deg r_1 < \deg g$, then $q = (a/b)x^{n-m}$ and $r = r_1$. Otherwise one applies the same reduction to r_1 and arrives, after a finite number of steps, to the desired presentation.

We would like to have a similar division algorithm for polynomials in several variables. In Section 2.3 we will give such an algorithm. Apparently there are two problems to overcome. The first is, that it is not clear what the leading term of such a polynomial should be. For example, what should be the leading term of $f = x_1^2x_2 + x_1x_2^2$? To define a leading term we must have a total order on the monomials. The second is, that one has to say what it means that the remainder is "small" compared with the divisor. We will deal with these problems in the following sections.

2.1. Monomial orders

2.1.1. Examples and basic properties of monomial orders. Let X be a set. A **partial order** on X is a binary relation $\leq$ over X which is

reflexive, **antisymmetric** and **transitive**, i.e., for all a, b, and c in X, we have:

- $a \leq a$ (reflexivity);
- if $a \leq b$ and $b \leq a$, then $a = b$ (antisymmetry);
- if $a \leq b$ and $b \leq c$, then $a \leq c$ (transitivity).

It is common to write $a < b$ if $a \leq b$ and $a \neq b$. We also write $a \geq b$ ($a > b$), if $b \leq a$ ($b < a$). A typical example of a partially ordered set is the set of all subsets of a given set ordered by inclusion.

A partial order $\leq$ on X is called a **total order**, if for any two elements $a, b \in X$ one has $a \leq b$ or $b \leq a$.

Let K be a field and $S = K[x_1, \ldots, x_n]$. We now define a total order on $\operatorname{Mon}(S)$, the set of all monomials in S, which respects the multiplicative structure on this set.

Definition 2.1. *A* ***monomial order*** *on S is a total order $\leq$ on $\operatorname{Mon}(S)$ with the properties:*

(i) *$1 \leq u$ for all $u \in \operatorname{Mon}(S)$;*
(ii) *if $u < v$ and $w \in \operatorname{Mon}(S)$, then $uw < vw$.*

A monomial order satisfies the following two conditions.

Proposition 2.2. *Let $\leq$ be a monomial order on S. Then the following holds:*

(i) *if $u, v \in \operatorname{Mon}(S)$ with $u|v$, then $u \leq v$;*
(ii) *if $u_1, u_2, \ldots$ is a sequence of monomials with $u_1 \geq u_2 \geq \cdots$, then there exists an integer m such that $u_i = u_m$ for all $i \geq m$.*

Proof. (i) If $u|v$, then there exists a monomial w such that $v = uw$. Since $1 \leq w$, it follows that $u \leq wu = v$.

(ii) Let $\mathcal{M} = \{u_1, u_2, \ldots\}$. By Dickson's lemma this set has, with respect to divisibility, only a finite number of minimal elements, say $u_{i_1}, \ldots, u_{i_r}$ with $i_1 < i_2 < \cdots < i_r$. Let j be any integer $\geq i_r$. Then there exists an integer $1 \leq k \leq r$ such that $u_{i_k}|u_j$. By (i), this implies that $u_{i_k} \leq u_j$. Hence $u_{i_k} \geq u_{i_r} \geq u_j \geq u_{i_k}$, and so $u_{i_r} = u_j$. Therefore we may choose $m = i_r$. $\square$

Next we consider some standard monomial orders on S. Let $\mathbf{x}^{\mathbf{a}}, \mathbf{x}^{\mathbf{b}}$ be two monomials in S.

The lexicographic order: $\mathbf{x}^{\mathbf{a}} < \mathbf{x}^{\mathbf{b}}$, if either $\sum_{i=1}^{n} a_i < \sum_{i=1}^{n} b_i$ or $\sum_{i=1}^{n} a_i = \sum_{i=1}^{n} b_i$, and the left-most nonzero component of $\mathbf{a} - \mathbf{b}$ is negative.

The pure lexicographic order: $\mathbf{x}^{\mathbf{a}} < \mathbf{x}^{\mathbf{b}}$, if the left-most nonzero component of $\mathbf{a} - \mathbf{b}$ is negative.

The reverse lexicographic order: $\mathbf{x}^{\mathbf{a}} < \mathbf{x}^{\mathbf{b}}$, if either $\sum_{i=1}^{n} a_i < \sum_{i=1}^{n} b_i$ or $\sum_{i=1}^{n} a_i = \sum_{i=1}^{n} b_i$, and the right-most nonzero component of $\mathbf{a} - \mathbf{b}$ is positive.

Observe that for each of these three orders we have $x_1 > x_2 > \cdots > x_n$.

The difference between the lexicographic and the reverse lexicographic order is subtle. Consider, for example, the monomials of degree 3 in 3 variables in lexicographic order,

$$x_1^3 > x_1^2x_2 > x_1^2x_3 > x_1x_2^2 > x_1x_2x_3 > x_1x_3^2 > x_2^3 > x_2^2x_3 > x_2x_3^2 > x_3^3,$$

and in reverse lexicographic order,

$$x_1^3 > x_1^2x_2 > x_1x_2^2 > x_2^3 > x_1^2x_3 > x_1x_2x_3 > x_2^2x_3 > x_1x_3^2 > x_2x_3^2 > x_3^3.$$

A good rule to memorize the lexicographic order and the reverse lexicographic order is the following: $u > v$ in the lexicographic order if u has "more from the beginning" than v; and $u > v$ in the reverse lexicographic order if u has "less from the end" than v.

2.1.2. Construction of monomial orders. We describe a few techniques to create new monomial orders from given ones. For the standard monomial orders described in the previous subsection we have $x_n < x_{n-1} < \cdots < x_1$. There are situations where one would like to have a different order of the variables. This is easily achieved: let $<$ be an arbitrary monomial order with $x_n < x_{n-1} < \cdots < x_1$ and let $\pi\colon [n] \to [n]$ be a permutation of the set of integers $[n] = \{1, \ldots, n\}$. We define a new monomial order $<_\pi$ as follows:

$$\mathbf{x}^{\mathbf{a}} <_\pi \mathbf{x}^{\mathbf{b}} \quad \Leftrightarrow \quad \mathbf{x}^{\pi(\mathbf{a})} < \mathbf{x}^{\pi(\mathbf{b})},$$

where for $\mathbf{c} = (c_1, \ldots, c_n) \in \mathbb{Z}^n$ we set $\pi(\mathbf{c}) = (c_{\pi(1)}, \ldots, c_{\pi(n)})$.

This new monomial order satisfies

$$x_{\pi(n)} <_\pi x_{\pi(n-1)} <_\pi \cdots <_\pi x_{\pi(1)}.$$

The pure lexicographic order may be viewed as a product order. For $i = 1, \ldots, r$ let $S_i = K[x_{i1}, \ldots, x_{is_i}]$ be a polynomial ring over K, and let

$$S = K[x_{11}, \ldots, x_{1s_1}, x_{21}, \ldots, x_{rs_r}]$$

be the polynomial ring over K in all the variables x_{ij}. Furthermore, let $<_i$ be a monomial order on S_i. We define the **product order** $<$ on S as follows: let $u, v \in \operatorname{Mon}(S)$. There are unique monomials $u_i, v_i \in S_i$ such that $u = u_1u_2\cdots u_r$ and $v = v_1v_2\cdots v_r$. We set

$$u < v \quad \Leftrightarrow \quad u_1 = v_1, \ldots, u_{k-1} = v_{k-1},\ u_k <_k v_k \quad \text{for some } k.$$

For example, let $S_1 = K[x_1, x_2]$, $S_2 = K[y_1, y_2, y_3]$, $S = K[x_1, x_2, y_1, y_2, y_3]$. Let $<_1$ be the lexicographic order on S_1, $<_2$ the reverse lexicographic order on S_2 and $<$ the product order of $<_1$ and $<_2$. Then

$$x_1 x_2^2 y_2^2 > x_1 x_2^2 y_1 y_3 > x_1 x_2 y_2^5.$$

Let $<$ be any monomial order. We choose a vector $\mathbf{w} = (w_1, \ldots, w_n) \in \mathbb{R}^n$ with nonnegative entries, called **weight vector**. Then we define the new monomial order $<_{\mathbf{w}}$ as follows:

$$\mathbf{x}^{\mathbf{a}} <_{\mathbf{w}} \mathbf{x}^{\mathbf{b}}, \quad \text{if} \quad \mathbf{a} \cdot \mathbf{w} < \mathbf{b} \cdot \mathbf{w}, \quad \text{or else} \quad \mathbf{a} \cdot \mathbf{w} = \mathbf{b} \cdot \mathbf{w} \quad \text{and} \quad \mathbf{x}^{\mathbf{a}} < \mathbf{x}^{\mathbf{b}}.$$

Here $\mathbf{c} \cdot \mathbf{d} = \sum_{i=1}^{n} c_i d_i$ is the standard scalar product on $\mathbb{R}^n$.

For example, if $<$ is the pure lexicographic order and we choose $\mathbf{w} = (1, 1, \ldots, 1)$ as weight vector, then $<_{\mathbf{w}}$ is the lexicographic order.

For later applications the following example is important.

Proposition 2.3. *Let $<$ be any monomial order on $K[x_1, \ldots, x_n]$. Fix an integer $1 \leq t \leq n$ and choose the weight vector $\mathbf{w} = (1, 1, \ldots, 1, 0, 0, \ldots, 0)$ with the first t entries being 1 and the remaining last entries being 0. Then the order $<_{\mathbf{w}}$ has the following property: if u, v are monomials such that $x_j | u$ for some $j \leq t$, and x_j does not divide v for any $j \leq t$, then $v <_{\mathbf{w}} u$.*

Proof. Let $u = \mathbf{x}^{\mathbf{a}}$ and $v = \mathbf{x}^{\mathbf{b}}$. Then $a_j \neq 0$ for some $j \leq t$, hence $\mathbf{a} \cdot \mathbf{w} > 0$, while on the other hand, $b_j = 0$ for all $j \leq t$. Therefore, $\mathbf{b} \cdot \mathbf{w} = 0$. This yields the desired conclusion. □

2.2. Initial ideals and Gröbner bases

2.2.1. The basic definitions. We now come to the main topic of this book. Let as before $S = K[x_1, \ldots, x_n]$ be the polynomial ring over the field K, and let $<$ be a monomial order on S.

If $f \neq 0$ is a polynomial in S, we set $\mathtt{in}_<(f)$ to be the largest monomial $u \in \operatorname{supp}(f)$ with respect to $<$, and call it the **initial monomial** of f. The coefficient c of $\mathtt{in}_<(f)$ in f is called the **leading coefficient** of f with respect to $<$, and $c\,\mathtt{in}_<(f)$ is called the **leading term** of f.

For convenience we set $\mathtt{in}_<(0) = 0$ and let $\mathtt{in}_<(0) < \mathtt{in}_<(f)$ for all $f \neq 0$.

For example, let $f = 2x_1^2 x_3 + 3x_1 x_2^2$. Then $\mathtt{in}_<(f) = x_1^2 x_3$, if $<$ is the lexicographic order, and $\mathtt{in}_<(f) = x_1 x_2^2$, if $<$ is the reverse lexicographic order.

For the initial monomial of a product or a sum of polynomials we have the following rules.

Lemma 2.4. *Let $<$ be a monomial order on S and let $f_1, \ldots, f_r$ be nonzero polynomials in S. Then*

(i) $\mathtt{in}_<(f_1 f_2 \cdots f_r) = \mathtt{in}_<(f_1)\,\mathtt{in}_<(f_2) \cdots \mathtt{in}_<(f_r)$;

(ii) $\mathtt{in}_<(f_1 + f_2 + \cdots + f_r) \leq \max\{\mathtt{in}_<(f_1), \mathtt{in}_<(f_2), \ldots, \mathtt{in}_<(f_r)\}$;

(iii) *Let c_j be the leading coefficient of f_j. Equality holds in* (ii) *if and only if $\sum_j c_j \neq 0$, where the sum is taken only over those j for which $\mathtt{in}_<(f_j) \geq \mathtt{in}_<(f_i)$ for all i.*

Proof. (i) Since for all $u \in \mathrm{supp}(f_i)$ we have $\mathtt{in}_<(f_i) \geq u$, it follows that

$$\mathtt{in}_<(f_1)\,\mathtt{in}_<(f_2) \cdots \mathtt{in}_<(f_r) \geq u_1 u_2 \cdots u_r$$

for all $u_i \in \mathrm{supp}(f_i)$. Equality holds if and only if $u_i = \mathtt{in}_<(f_i)$ for $i = 1, \ldots, r$. Since all monomials in $\mathrm{supp}(f_1 f_2 \cdots f_r)$ are of the form $u_1 u_2 \cdots u_r$ with $u_i \in \mathrm{supp}(f_i)$, and since $\mathtt{in}_<(f_1)\,\mathtt{in}_<(f_2) \cdots \mathtt{in}_<(f_r)$ belongs to the support of $f_1 f_2 \cdots f_r$, the assertion follows.

(ii) Observe that $\mathrm{supp}(f_1 + \cdots + f_r) \subseteq \bigcup_{i=1}^r \mathrm{supp}(f_i)$. This implies that

$$\begin{aligned} \mathtt{in}_<(f_1 + \cdots + f_r) &\leq \max\{u \,:\, u \in \bigcup_{i=1}^r \mathrm{supp}(f_i)\} \\ &= \max\{\mathtt{in}_<(f_1), \mathtt{in}_<(f_2), \ldots, \mathtt{in}_<(f_r)\}. \end{aligned}$$

(iii) Let u be the maximal initial monomial appearing among the f_i. Assuming that $\sum_j c_j \neq 0$, where the sum is taken over those j for which $\mathtt{in}_<(f_j) = u$, it follows that $u \in \mathrm{supp}(f_1 + \cdots + f_r)$. Therefore,

$$\mathtt{in}_<(f_1 + \cdots + f_r) \geq u = \max\{\mathtt{in}_<(f_1), \mathtt{in}_<(f_2), \ldots, \mathtt{in}_<(f_r)\}.$$

Thus, by (ii), equality holds.

Conversely, if $\sum_{j=1}^k c_j = 0$, then $u \notin \mathrm{supp}(f_1 + f_2 + \cdots + f_r)$, and hence $\mathtt{in}_<(f_1 + f_2 + \cdots + f_r) \neq \max\{\mathtt{in}_<(f_1), \mathtt{in}_<(f_2), \ldots, \mathtt{in}_<(f_r)\}$. ☐

Let $I \subset S$ be a nonzero ideal. The **initial ideal** of I is the monomial ideal

$$\mathtt{in}_<(I) = (\mathtt{in}_<(f) \,:\, f \in I,\ f \neq 0).$$

If $I = (0)$, then we set $\mathtt{in}_<(I) = (0)$.

Notice that $\mathtt{in}_<(I)$ is generated by the initial monomials of *all* nonzero polynomials in I. In general the initial monomials of a set of generators do *not* generate $\mathtt{in}_<(I)$. Consider, for example, the ideal

$$I = (f, g) \quad \text{with} \quad f = x_1x_2 - x_3x_4,\ g = -x_2^2 + x_1x_3.$$

With respect to the reverse lexicographic order $<$ we have $\mathtt{in}_<(f) = x_1x_2$ and $\mathtt{in}_<(g) = x_2^2$. On the other hand, $h = x_2 f + x_1 g = x_1^2x_3 - x_2x_3x_4 \in I$ and $\mathtt{in}_<(h) = x_1^2x_3$. Thus we see that $\mathtt{in}_<(h) \notin (\mathtt{in}_<(f), \mathtt{in}_<(g))$.

A priori $\mathtt{in}_<(I)$ is generated by infinitely many initial monomials. Nevertheless, since $\mathtt{in}_<(I)$ is a monomial ideal, as we know from Corollary 1.10, there exists $g_1, \ldots, g_m \in I$ such that

$$\mathtt{in}_<(I) = (\mathtt{in}_<(g_1), \ldots, \mathtt{in}_<(g_m)).$$

The observation leads to the most important concept studied in this book.

Definition 2.5. *Let $I \subset S$ be an ideal, and let $<$ be a monomial order on S. A sequence $g_1, \ldots, g_m$ of elements in I with $\mathtt{in}_<(I) = (\mathtt{in}_<(g_1), \ldots, \mathtt{in}_<(g_m))$ is called a* ***Gröbner basis*** *of I with respect to the monomial order $<$.*

The argument that we used to see that a Gröbner basis of I always exist does not tell us how to actually find a Gröbner basis. In case that I is a principal ideal, say $I = (g)$, one sees immediately that g is a Gröbner basis of I.

The Buchberger algorithm, which will be discussed in the next chapter, gives an efficient method to compute a Gröbner basis of an ideal. For many abstract arguments, however, it just suffices to know that a Gröbner basis always exists. First remarkable examples of such reasoning are the next theorems.

2.2.2. Macaulay's theorem. The following theorem is fundamental in Gröbner basis theory and it will be used several times later in the book.

Theorem 2.6 (Macaulay). *Let $<$ be a monomial order on S, and let $I \subset S$ an ideal. Then the monomials in S which do not belong to $\mathtt{in}_<(I)$ form a K-basis of S/I.*

Proof. Suppose the monomials in S which do not belong to $\mathtt{in}_<(I)$ are K-linearly dependent modulo I. Then there exists a nonzero polynomial $f \in I$ with $\mathrm{supp}(f) \cap \mathrm{Mon}(\mathtt{in}_<(I)) = \emptyset$, where $\mathrm{Mon}(\mathtt{in}_<(I))$ denotes the set of monomials in $\mathtt{in}_<(I)$. This contradicts the fact that $\mathtt{in}_<(f) \in \mathtt{in}_<(I)$.

It remains to be shown that residue classes of the monomials in S which do not belong to $\mathtt{in}_<(I)$ generate the K-vector space S/I. To this end we will show that for any $f \in S$ there exists $g \in S$ with $f + I = g + I$ and $\mathrm{supp}(g) \cap \mathrm{Mon}(\mathtt{in}_<(I)) = \emptyset$. Suppose this is not the case, and let $f \in S$ be a polynomial with smallest initial monomial for which we cannot find a polynomial g as above. Let c be the leading coefficient of f. Then $f - c\,\mathtt{in}_<(f)$ has a smaller initial monomial, and hence there exists $g \in S$ with $\mathrm{supp}(g) \cap \mathrm{Mon}(\mathtt{in}_<(I)) = \emptyset$ and such that $(f - c\,\mathtt{in}_<(f)) + I = g + I$. Thus $f + I = (c\,\mathtt{in}_<(f) + g) + I$. If $\mathtt{in}_<(f) \notin \mathtt{in}_<(I)$, we may replace g by $c\,\mathtt{in}_<(f) + g$, contradicting the choice of f. If $\mathtt{in}_<(f) \in \mathtt{in}_<(I)$, then there exists $h \in I$ with leading coefficient 1 and $\mathtt{in}_<(h) = \mathtt{in}_<(f)$. It follows that $f - ch$ has a smaller initial monomial than f. Thus we can find a polynomial

$g \in S$ with $\text{supp}(g) \cap \text{Mon}(\texttt{in}_<(I)) = \emptyset$ and $(f - ch) + I = g + I$. Since $f + I = (f - ch) + I$, this contradicts again the choice of f. $\square$

On a polynomial ring with more than one variable there exist infinitely many different monomial orders; see Problem 2.4. However, as an application of Macaulay's theorem we show

Proposition 2.7. *Every ideal $I \subset S$ has only finitely many distinct initial ideals.*

Proof. Let $I \subset S$ be an ideal, and let $\mathcal{S}_0$ be the set of initial ideals of I. Assume that $\mathcal{S}_0$ is an infinite set. Let $0 \neq f_1 \in I$. To each initial ideal $J \in \mathcal{S}_0$ belongs a monomial $u \in \text{supp}(f_1)$ with $u \in J$. Since $\text{supp}(f_1)$ is a finite set, there exists $u_1 \in \text{supp}(f_1)$ such that the set $\mathcal{S}_1 = \{J \in \mathcal{S}_0 \colon\ u_1 \in J\}$ is infinite. In particular, $J \neq (u_1)$ for at least one (in fact, infinitely many) $J \in \mathcal{S}_1$. Thus Theorem 2.6 implies that the monomials which do not belong to (u_1) are linearly dependent modulo I. Hence there exists $0 \neq f_2 \in I$ with $\text{supp}(f_2) \cap (u_1) = \emptyset$. As before there exists $u_2 \in \text{supp}(f_2)$ such that $\mathcal{S}_2 = \{J \in \mathcal{S}_1 \colon\ u_2 \in J\}$ is an infinite set. Since $u_2 \notin (u_1)$ it follows that (u_1) is strictly contained in (u_1, u_2). Again, since $\mathcal{S}_2$ is infinite, $J \neq (u_1, u_2)$ for some $J \in \mathcal{S}_2$, and as before we can construct an element $u_3 \notin (u_1, u_2)$. Proceeding in this way we construct an infinite strictly ascending sequence $(u_1) \subset (u_1, u_2) \subset (u_1, u_2, u_3) \subset \cdots$ of monomial ideals, contradicting Proposition 1.12. $\square$

2.2.3. Hilbert's basis theorem. In the previous section we have seen that a system of generators of an ideal I need not to be a Gröbner basis of I. However, we have

Theorem 2.8. *Let $I \subset S$ be an ideal and let $g_1, \dots, g_m$ be a Gröbner basis of I with respect to a monomial order $<$. Then $g_1, \dots, g_m$ is a system of generators of I.*

Proof. If $f \in I$, then $\texttt{in}_<(f) \in \texttt{in}_<(I) = (\texttt{in}_<(g_1), \dots, \texttt{in}_<(g_m))$. Therefore there exists an integer $1 \leq i \leq m$ and a monomial w such that $\texttt{in}_<(f) = w\,\texttt{in}_<(g_i)$. Let c be the coefficient of $\texttt{in}_<(f)$ in f and d the coefficient of $\texttt{in}_<(g_i)$ in g_i, and let $h = f - cd^{-1}wg_i$. Then $h \in I$. If $h = 0$, then $f = cd^{-1}wg_i \in (g_1, \dots, g_m)$. Now assume $h \neq 0$. Since $\texttt{in}_<(f) > \texttt{in}_<(h)$ and since by Proposition 2.2 each strictly descending chain of monomials terminates, we may apply an induction argument and hence may assume that h is a linear combination of the g_i with coefficients in S. Since $f = h + cd^{-1}wg_i$, the same is true for f. $\square$

Since each ideal has a Gröbner basis, and each Gröbner basis contains finitely many elements, we get

Corollary 2.9 (Hilbert's basis theorem). *Each ideal in the polynomial ring $S = K[x_1, \dots, x_n]$ is finitely generated.*

Hilbert's basis theorem says that the polynomial ring $S = K[x_1, \dots, x_n]$ is Noetherian. It is known from general Commutative Algebra that a Noetherian ring is also characterized by the property that each ascending chain of ideals terminates. We will give a direct proof of this property for polynomial rings.

Proposition 2.10. *Let $I_1 \subseteq I_2 \subseteq \cdots$ be an ascending chain of ideals in S. Then there exists an integer k such that $I_j = I_k$ for all $j \geq k$.*

Proof. Fix some monomial order $<$ on S. The given chain of ideals induces the following chain $\mathtt{in}_<(I_1) \subseteq \mathtt{in}_<(I_2) \subseteq \cdots$ of monomial ideals. By Proposition 1.12, there exists an integer k such that $\mathtt{in}_<(I_j) = \mathtt{in}_<(I_k)$ for all $j \geq k$. It follows from Problem 2.8 that $I_j = I_k$ for all $j \geq k$. □

2.3. The division algorithm

As announced in the introduction of this chapter we now discuss an extension of the classical division algorithm. The extension will be two-fold. On the one hand, the polynomials in one variable will be replaced by polynomials in several variables, and on the other hand, we may "divide" not only by one but if we wish by several polynomials at the same time. The precise statement is formulated in the next

Theorem 2.11. *Let f and $g_1, \dots, g_m$ be polynomials in S with $g_i \neq 0$. Given a monomial order $<$, there exist polynomials $q_1, \dots, q_m$ and a polynomial r in S with*

$$f = q_1 g_1 + q_2 g_2 + \cdots + q_m g_m + r$$

such that the following conditions are satisfied:

(i) *no element of* $\mathrm{supp}(r)$ *is contained in the ideal* $(\mathtt{in}_<(g_1), \dots, \mathtt{in}_<(g_m))$;

(ii) $\mathtt{in}_<(f) \geq \mathtt{in}_<(q_i g_i)$ *for all i.*

An equation $f = q_1 g_1 + q_2 g_2 + \cdots + q_m g_m + r$ satisfying the conditions (i) and (ii) is called a **standard expression** of f, and r is called a **remainder** of f with respect to $g_1, \dots, g_m$. The polynomial f may have different standard expressions and different remainders with respect to $g_1, \dots, g_m$ as the following example demonstrates. Let $f = x_1 x_2 + x_2^2$, $g_1 = x_1 + x_2$ and $g_2 = x_1$. We let $<$ be the lexicographic order. Then

$$f = x_2 g_1 \quad \text{as well as} \quad f = x_2 g_2 + x_2^2$$

are standard expressions of f. In the first case the remainder is 0, in the second case the remainder is x_2^2.

We say that f **reduces to** 0 with respect to $g_1, \dots, g_m$, if f has a remainder with respect to $g_1, \dots, g_m$ which is 0.

Proof of Theorem 2.11. If no element of $\operatorname{supp}(f)$ is contained in the ideal $(\mathtt{in}_<(g_1), \dots, \mathtt{in}_<(g_m))$, we may choose $q_i = 0$ for $i = 1, \dots, m$ and $r = f$. Now suppose that some $u \in \operatorname{supp}(f)$ belongs to $(\mathtt{in}_<(g_1), \dots, \mathtt{in}_<(g_m))$. Let u_0 be the largest such monomial in $\operatorname{supp}(f)$. Then $\mathtt{in}_<(g_j)$ divides u_0 for some j. Thus $u_0 = w\,\mathtt{in}_<(g_j)$ for some monomial w in S. Let c be the coefficient of u_0 in f and d the coefficient of $\mathtt{in}_<(g_j)$ in g_j, and set

$$h = f - cd^{-1}wg_j. \tag{2.1}$$

If $h = 0$ or no $u \in \operatorname{supp}(h)$ belongs to $(\mathtt{in}_<(g_1), \dots, \mathtt{in}_<(g_m))$, then $f = cd^{-1}wg_j + h$ is a standard expression for f with remainder $r = h$. Otherwise there exists $u \in \operatorname{supp}(h)$ which belongs to $(\mathtt{in}_<(g_1), \dots, \mathtt{in}_<(g_m))$. Let u_1 be the largest such monomial in $\operatorname{supp}(h)$. We claim that $u_1 < u_0$. In fact, the definition of h implies $u_0 \notin \operatorname{supp}(h)$ and that $u_1 \in \operatorname{supp}(f) \cup \operatorname{supp}(wg_j)$. If $u_1 \in \operatorname{supp}(f)$, then by the choice of u_0 it follows that $u_1 < u_0$. On the other hand, if $u_1 \in \operatorname{supp}(wg_j)$, then $u_1 \leq \mathtt{in}_<(wg_j) = u_0$. Since $u_1 \neq u_0$ it follows in this case as well that $u_1 < u_0$. Thus, since the maximal monomial in the support of h which does belong to $(\mathtt{in}_<(g_1), \dots, \mathtt{in}_<(g_m))$ is smaller than that of f and since descending sequences of monomials terminate by Proposition 2.2, we may assume by induction that h has a standard presentation $h = \sum_{i=1}^m \tilde{q}_i g_i + r$. It follows that $f = q_1g_1 + q_2g_2 + \cdots + q_mg_m + r$ with $q_j = \tilde{q}_j + cd^{-1}w$ and $q_i = \tilde{q}_i$ for $i \neq j$ is a standard presentation of f. Indeed, since $h = \sum_{j=1}^m \tilde{q}_j g_j + r$ is a standard expression for h condition (i) is satisfied for the expression $f = \sum_{i=1}^m q_i g_i + r$. To see that condition (ii) is also satisfied for this expression of f we notice that due to (2.1) and in view of Lemma 2.4 we have

$$\mathtt{in}_<(h) \leq \max\{\mathtt{in}_<(f), u_0\} = \mathtt{in}_<(f). \tag{2.2}$$

Hence since $\mathtt{in}_<(h) \geq \mathtt{in}_<(\tilde{q}_i g_i)$ for all i, it follows from (2.2) that $\mathtt{in}_<(f) \geq \mathtt{in}_<(q_i g_i)$ for all $i \neq j$. Finally, since $\mathtt{in}_<(q_j g_j) \leq \max\{\mathtt{in}_<(\tilde{q}_j g_j), u_0\}$, we see that $\mathtt{in}_<(f) \geq \mathtt{in}_<(q_j g_j)$ as well, since $\mathtt{in}_<(f) \geq \mathtt{in}_<(h) \geq \mathtt{in}_<(\tilde{q}_j g_j)$ and since $\mathtt{in}_<(f) \geq u_0$. □

Our proof of Theorem 2.11 provides an algorithm to compute a standard expression of f with respect to $g_1, \dots, g_m$. For the algorithm which we now describe, we fix the order of the elements $g_1, \dots, g_m$. Then the remainder resulting from this algorithm will be uniquely determined.

The algorithm produces a finite sequence of polynomials h_i as follows: $h_0 = f$. Suppose $h_0, \dots, h_i$ is already defined. The sequence ends with h_i if no $u \in \operatorname{supp}(h_i)$ belongs to $(\mathtt{in}_<(g_1), \dots, \mathtt{in}_<(g_m))$. Otherwise, let u be the largest monomial in the support of h_i which belongs to $(\mathtt{in}_<(g_1),\ldots,\mathtt{in}_<(g_m))$ and let j be the smallest integer such that $\mathtt{in}_<(g_j)|u$. We define

$$h_{i+1} = h_i - cd^{-1}wg_j,$$

where $w = u/\,\mathtt{in}_<(g_j)$ and where c is the leading coefficient of h_i and d the leading coefficient of g_j.

Say, the sequence of the h_i ends with h_s. Then we have the following set of equations:

$$f = h_0 = \tilde{q}_1 g_{j_1} + h_1, \tag{2.3}$$

$$h_1 = \tilde{q}_2 g_{j_2} + h_2, \tag{2.4}$$

$$h_2 = \tilde{q}_3 g_{j_3} + h_3, \tag{2.5}$$

$$\vdots \qquad \vdots$$

$$h_{s-1} = \tilde{q}_s g_{j_s} + h_s. \tag{2.6}$$

Replacing h_1 in (2.3) by the expression of h_1 given in (2.4), we obtain $f = \tilde{q}_1 g_{j_1} + \tilde{q}_2 g_{j_2} + h_2$. Now we replace h_2 in this new expression of f by the expression of h_2 given in (2.5). Continuing this way we arrive at the standard expression

$$f = q_1 g_1 + q_2 g_2 + \cdots + q_m g_m + r,$$

where $r = h_s$ and $q_i = \sum_{k, j_k = i} \tilde{q}_k$.

We demonstrate this algorithm with the following example: let $f = x_1^3 - x_1^2 x_2 + 2x_1 x_2^2 + x_1 x_2 x_3$, $g_1 = x_1^2 - 1$ and $g_2 = x_1 x_2 - x_2 x_3$. We compute a standard expression of f with respect to g_1, g_2 for the lexicographic order, and get

$$\begin{aligned} f = h_0 &= x_1 g_1 + h_1, & h_1 &= -x_1^2 x_2 + 2x_1 x_2^2 + x_1 x_2 x_3 + x_1, \\ h_1 &= -x_2 g_1 + h_2, & h_2 &= 2x_1 x_2^2 + x_1 x_2 x_3 + x_1 - x_2, \\ h_2 &= 2x_2 g_2 + h_3, & h_3 &= x_1 x_2 x_3 + 2x_2^2 x_3 + x_1 - x_2, \\ h_3 &= x_3 g_2 + h_4, & h_4 &= 2x_2^2 x_3 + x_2 x_3^2 + x_1 - x_2. \end{aligned}$$

Thus we obtain the standard expression

$$\begin{aligned} f &= x_1 g_1 + h_1 \\ &= x_1 g_1 + (-x_2 g_1 + h_2) \\ &= (x_1 - x_2) g_1 + (2x_2 g_2 + h_3) \\ &= (x_1 - x_2) g_1 + 2x_2 g_2 + (x_3 g_2 + h_4) \\ &= (x_1 - x_2) g_1 + (2x_2 + x_3) g_2 + (2x_2^2 x_3 + x_2 x_3^2 + x_1 - x_2). \end{aligned}$$

Even though, as we have seen, a polynomial may have different standard expressions, one has the following uniqueness statement.

Proposition 2.12. *Given a monomial order $<$ on S. Assume that the polynomials $g_1, \ldots, g_m$ form a Gröbner basis of the ideal $I = (g_1, \ldots, g_m)$. Then each polynomial $f \in S$ has a unique remainder with respect to $g_1, \ldots, g_m$.*

Proof. Let $f \in S$, and let $f = \sum_{i=1}^m q_i g_i + r$ and $f = \sum_{i=1}^m p_i g_i + s$ be two standard expressions for f. Then $h = r - s \in I$. Suppose that $h \neq 0$. Then $\mathtt{in}_<(h) \in \mathtt{in}_<(I) = (\mathtt{in}_<(g_1), \ldots, \mathtt{in}_<(g_m))$. Since $\mathtt{in}_<(h) \in \operatorname{supp}(r) \cup \operatorname{supp}(s)$ and r, s are remainders of f, we arrive at a contradiction. □

Proposition 2.12 has the following important consequence.

Corollary 2.13. *Let $I = (g_1, \ldots, g_m) \subset S$ be an ideal and assume that for some monomial order, $g_1, \ldots, g_m$ is a Gröbner basis of I. Then a polynomial $f \in S$ belongs to I if and only if f reduces to 0 with respect to $g_1, \ldots, g_m$.*

Proof. Obviously, f belongs to I if the remainder of f with respect to $g_1, \ldots, g_m$ is 0. Conversely, assume that $f \in I$ and let $f = q_1 g_1 + \cdots + q_m g_m + r$ be a standard expression of f. Then $r \in I$. Suppose that $r \neq 0$. Then $\mathtt{in}_<(r) \in \mathtt{in}_<(I) = (\mathtt{in}_<(g_1), \ldots, \mathtt{in}_<(g_m))$, contradicting the definition of a remainder. □

2.4. Buchberger's criterion

So far we have no method to check whether a system of generators of an ideal is a Gröbner basis with respect to some monomial order $<$. We will now derive a criterion which allows us to answer this question in a finite number of steps. To explain this criterion we have to introduce the so-called S-polynomials. Suppose first we are dealing with an ideal generated by two nonzero polynomials, say $I = (f, g)$, and we want to compute a Gröbner basis of I. Certainly $\mathtt{in}_<(f)$ and $\mathtt{in}_<(g)$ belong to $\mathtt{in}_<(I)$. A candidate of a polynomial $h \in I$ whose initial monomial does not belong to $(\mathtt{in}_<(f), \mathtt{in}_<(g))$ is a linear combination of f and g such that their initial terms cancel. This leads us to define

$$S(f,g) = \frac{\operatorname{lcm}(\mathtt{in}_<(f), \mathtt{in}_<(g))}{c\,\mathtt{in}_<(f)} f - \frac{\operatorname{lcm}(\mathtt{in}_<(f), \mathtt{in}_<(g))}{d\,\mathtt{in}_<(g)} g,$$

where c is the leading coefficient of f and d is the leading coefficient of g. The polynomial $S(f,g)$ is called the **S-polynomial** of f and g with respect to $<$.

Now we are ready to formulate the celebrated **Buchberger criterion**.

Theorem 2.14. *Let $<$ be a monomial order on S, and let $I = (g_1, \ldots, g_m)$ be an ideal in S with $g_i \neq 0$ for all i. Then the following conditions are equivalent:*

(a) *$g_1, \ldots, g_m$ is a Gröbner basis of I with respect to $<$;*

(b) *$S(g_i, g_j)$ reduces to 0 with respect to $g_1, \ldots, g_m$ for all $i < j$.*

Proof. (a) $\Rightarrow$ (b): Since $S(g_i, g_j) \in I$ and since $g_1, \ldots, g_m$ is a Gröbner basis of I, it follows from Corollary 2.13 that the remainder of $S(g_i, g_j)$ with respect to $g_1, \ldots, g_m$ is 0.

(b) $\Rightarrow$ (a): We have to show that $\mathtt{in}_<(I) = (\mathtt{in}_<(g_1), \ldots, \mathtt{in}_<(g_m))$. In other words, if $0 \neq f \in I$, we need to see that the monomial $\mathtt{in}_<(f)$ belongs to $(\mathtt{in}_<(g_1), \ldots, \mathtt{in}_<(g_m))$. Since $f \in I$, there exist $h_1, \ldots, h_m \in S$ such that $f = \sum_{i=1}^m h_i g_i$. It follows from Lemma 2.4 that $\mathtt{in}_<(f) \leq \max_i\{\mathtt{in}_<(h_i g_i)\}$. If equality holds, then

$$\mathtt{in}_<(f) = \mathtt{in}_<(h_i g_i) = \mathtt{in}_<(h_i)\,\mathtt{in}_<(g_i)$$

for some i, and hence $\mathtt{in}_<(f) \in (\mathtt{in}_<(g_1), \ldots, \mathtt{in}_<(g_m))$. Otherwise, we have $\mathtt{in}_<(f) < \max_i\{\mathtt{in}_<(h_i g_i)\}$. We claim that in this case there exists a new presentation $f = \sum_{i=1}^m h_i' g_i$ of f with the property that $\max_i\{\mathtt{in}_<(h_i' g_i)\} < \max_i\{\mathtt{in}_<(h_i g_i)\}$. Then, if necessary we can modify the presentation of f again and again, until after a finite number of modifications, we obtain a presentation $f = \sum_{i=1}^m h_i'' g_i$ with $\mathtt{in}_<(f) = \max_i\{\mathtt{in}_<(h_i'' g_i)\}$, and we are done.

Let $u = \max_i\{\mathtt{in}_<(h_i g_i)\}$. In order to prove the claim, we may assume for simplicity that $\mathtt{in}_<(h_j g_j) = u$ for $j = 1, \ldots r$ and that $\mathtt{in}_<(h_j g_j) < u$ for $j > r$. We may also assume that all leading coefficients of the g_i are equal to 1. Let $w_j = \mathtt{in}_<(h_j)$ and c_j the leading coefficient of h_j. Then, since $\mathtt{in}_<(f) < \max_i\{\mathtt{in}_<(h_i g_i)\}$, it follows from Lemma 2.4 that $\sum_{j=1}^r c_j = 0$.

Since $u = w_1\,\mathtt{in}_<(g_1) = w_j\,\mathtt{in}_<(g_j)$ for $j = 1, \ldots, r$, we see that the monomial $\operatorname{lcm}(\mathtt{in}_<(g_1), \mathtt{in}_<(g_j))$ divides u. Let

$$v_j = \frac{w_1\,\mathtt{in}_<(g_1)}{\operatorname{lcm}(\mathtt{in}_<(g_1), \mathtt{in}_<(g_j))} = \frac{w_j\,\mathtt{in}_<(g_j)}{\operatorname{lcm}(\mathtt{in}_<(g_1), \mathtt{in}_<(g_j))}.$$

Then $v_j S(g_1, g_j) = w_1 g_1 - w_j g_j$ and $u > \mathtt{in}_<(v_j S(g_1, g_j))$. Since each $S(g_1, g_j)$ has a remainder which is 0, we find for all j an expression $S(g_1, g_j) = \sum_{i=1}^m h_{ji}' g_i$ with $h_{ji}' \in S$ and such that $\mathtt{in}_< S(g_1, g_j) \geq \mathtt{in}_<(h_{ji}' g_i)$ for all i with $h_{ji}' \neq 0$. It follows that $v_j S(g_1, g_j) = \sum_{i=1}^m v_j h_{ji}' g_i = \sum_{i=1}^m h_{ji} g_i$ with $h_{ji} = v_j h_{ji}'$ and that

$$u > \mathtt{in}_<(v_j S(g_1, g_j)) \geq \mathtt{in}_<(h_{ji} g_i) \quad \text{for all } j \text{ and } i.$$

Hence we have

$$w_1 g_1 - w_j g_j - \sum_{i=1}^m h_{ji} g_i = 0 \quad \text{and} \quad \mathtt{in}_<(h_{ji} g_i) < u \quad \text{for all } j \text{ and } i.$$

Keeping in mind that $\sum_{j=1}^{r} c_j = 0$, we have

$$\sum_{j=2}^{r} c_j(w_1 g_1 - w_j g_j) = -\sum_{j=1}^{r} c_j w_j g_j,$$

and we can rewrite f as follows:

$$\begin{aligned} f &= \sum_{i=1}^{r} h_i g_i + \sum_{i=r+1}^{m} h_i g_i \\ &= \sum_{i=1}^{r} h_i g_i + \sum_{j=2}^{r} c_j(w_1 g_1 - w_j g_j - \sum_{i=1}^{m} h_{ji} g_i) + \sum_{i=r+1}^{m} h_i g_i \\ &= \sum_{i=1}^{r} (h_i - c_i w_i) g_i + \sum_{i=r+1}^{m} h_i g_i - \sum_{i=1}^{m} (\sum_{j=2}^{r} c_j h_{ji}) g_i \\ &= \sum_{i=1}^{m} h'_i g_i, \end{aligned}$$

where

$$h'_i = \begin{cases} h_i - c_i w_i - \sum_{j=2}^{r} c_j h_{ji}, & \text{for } i = 1, \ldots, r, \\ h_i - \sum_{j=2}^{r} c_j h_{ji}, & \text{for } i = r+1, \ldots, m. \end{cases}$$

From the definition of the h'_i we get $\max_i\{\mathtt{in}_<(h'_i g_i)\} < \max_i\{\mathtt{in}_<(h_i g_i)\}$, as desired. □

Checking whether a system of generators $g_1, \ldots, g_m$ of an ideal is a Gröbner basis can be rather cumbersome since one has to compute the remainder of $\binom{m}{2}$ S-polynomials. The following result can sometimes be used to shorten the calculations significantly.

Proposition 2.15. *Let $<$ be a monomial order on S, and let $f, g \in S$ be two polynomials such that $\mathtt{in}_<(f)$ and $\mathtt{in}_<(g)$ are relatively prime. Then $S(f, g)$ reduces to 0 with respect to f, g.*

Proof. We may assume that the leading coefficients of f and g are equal to 1. Then $f = \mathtt{in}_<(f) + f_1$, $g = \mathtt{in}_<(g) + g_1$ and

$$S(f, g) = \mathtt{in}_<(g)f - \mathtt{in}_<(f)g = (g - g_1)f - (f - f_1)g = f_1 g - g_1 f.$$

We claim that $f_1 g - g_1 f$ is the standard expression for $S(f, g)$ (which implies that $S(f, g)$ has a remainder which is 0). Indeed, suppose that $\mathtt{in}_<(f_1 g) = \mathtt{in}_<(g_1 f)$. Then $\mathtt{in}_<(f_1)\,\mathtt{in}_<(g) = \mathtt{in}_<(g_1)\,\mathtt{in}_<(f)$, and hence, since $\mathtt{in}_<(f)$ and $\mathtt{in}_<(g)$ are relatively prime, it follows that $\mathtt{in}_<(f)$ divides $\mathtt{in}_<(f_1)$ which is a contradiction, since $\mathtt{in}_<(f_1) < \mathtt{in}_<(f)$. Thus we may now assume, without loss of generality, that $\mathtt{in}_<(f_1 g) > \mathtt{in}_<(g_1 f)$. Then Lemma 2.4

implies that $\mathtt{in}(f) = \mathtt{in}_<(f_1 g) > \mathtt{in}_<(g_1 f)$. This shows that $f_1 g - g_1 f$ is in fact a standard expression for $S(f, g)$. □

As an example of how to use the Buchberger criterion we consider the ideal I generated by all the two by two minors of the matrix

$$\begin{pmatrix} x_1 & x_2 & \dots & x_n \\ y_1 & y_2 & \dots & y_n \end{pmatrix}.$$

In other words, I is generated by the binomials $f_{ij} = x_i y_j - x_j y_i$, $1 \le i < j \le n$. We claim that this set of generators is a Gröbner basis of I with respect to the lexicographic order $<$ with $x_1 > x_2 > \cdots > x_n > y_1 > y_2 > \cdots > y_n$. We have $\mathtt{in}_<(f_{ij}) = x_i y_j$ for all i and j. We will show that all S-polynomials $S(f_{ij}, f_{kl})$ with $\{i, j\} \neq \{k, l\}$ have a remainder which is 0. If $i \neq k$ and $j \neq l$, then $\mathtt{in}_<(f_{ij})$ and $\mathtt{in}_<(f_{kl})$ are relatively prime, so that by Proposition 2.15 $S(f_{ij}, f_{kl})$ has remainder 0. If $i = k$, we may assume that $j < l$ and get

$$S(f_{ij}, f_{kl}) = S(f_{ij}, f_{il}) = -x_j y_i y_l + x_l y_i y_j = -y_i f_{jl}.$$

This is a standard expression of $S(f_{ij}, f_{kl})$ with remainder 0.

Finally, if $j = l$, we assume $i < k$ and get

$$S(f_{ij}, f_{kl}) = S(f_{ij}, f_{kj}) = x_i x_j y_k - x_k x_j y_i = x_j f_{ik},$$

which is again a standard expression of $S(f_{ij}, f_{kl})$ with remainder 0.

2.5. Buchberger's algorithm

Let $I \subset S$ be an ideal. In this section we discuss an algorithm that allows, starting from a finite system of generators $\mathcal{G}$ of I, to compute a Gröbner basis of I. The algorithm, called the **Buchberger algorithm** , which is a simple consequence of Theorem 2.14, is the following:

Step 1: For each pair of distinct elements of $\mathcal{G}$ we compute the S-polynomial and a remainder of it.

Step 2: If all S-polynomials reduce to 0, then the algorithm ends and $\mathcal{G}$ is a Gröbner basis of I (according to Theorem 2.14). Otherwise we add one of the nonzero remainders to our system of generators, call this new system of generators again $\mathcal{G}$ and go back to Step 1.

This algorithm ends in a finite number of steps. Indeed, each time when we add a nonzero remainder of an S-polynomial to $\mathcal{G}$, as described in the algorithm, the ideal $(\mathtt{in}_<(g) : g \in \mathcal{G})$ becomes strictly larger. This is implied by the definition of the remainder. Since by Proposition 1.12 any strictly ascending sequence of monomial ideals is finite, the assertion follows.

To demonstrate the algorithm, we consider again our example from Section 2.2:

$$I = (f, g) \quad \text{with} \quad f = x_1x_2 - x_3x_4, \; g = -x_2^2 + x_1x_3.$$

We want to compute a Gröbner basis of I with respect to the reverse lexicographic order. We begin with the given system of generators $\mathcal{G} = \{f, g\}$ of I, and compute the S-polynomial of f and g:

$$S(f, g) = x_2f + x_1g = x_1^2x_3 - x_2x_3x_4.$$

In this case the remainder h of $S(f, g)$ is equal to $S(f, g)$ itself. Thus we add $h = -x_2x_3x_4 + x_1^2x_3$ to $\mathcal{G}$. We repeat the steps with the new set of generators $\mathcal{G}$ of I, and hence have to determine the remainders of the S-polynomials $S(f, g)$, $S(f, h)$ and $S(g, h)$ with respect to f, g, h. Note that since we added h to the set of generators of I, we only need to compute the remainders of $S(f, h)$ and $S(g, h)$ with respect to f, g, h since $S(f, g)$ has the standard expression $S(f, g) = h$, thus the remainder of $S(f, g)$ is zero. Computing the standard expressions of $S(f, h)$ and $S(g, h)$, we find

$$S(f, h) = -x_3x_4g \quad \text{and} \quad S(g, h) = x_1x_3h + x_2x_3x_4g.$$

Thus we see that all remainders are 0. The algorithm ends and f, g, h is a Gröbner basis of I.

2.6. Reduced Gröbner bases

Let $<$ be a monomial order on $S = K[x_1, \ldots, x_n]$, and $I \subset S$ an ideal. Then a Gröbner basis of I with respect to $<$ is of course not uniquely determined. For example to any Gröbner basis $\mathcal{G}$ of I one could add a few more elements of I to $\mathcal{G}$ and would obtain another Gröbner basis. Thus only with some extra conditions can a Gröbner basis be unique.

Definition 2.16. *Let $I \subset S$ be an ideal in S. Then $\mathcal{G} = g_1, \ldots, g_m$ is called a* ***reduced*** *Gröbner basis of I with respect to $<$, if $\mathcal{G}$ is a Gröbner basis of I with respect to the monomial order $<$ satisfying the following conditions:*

(i) *the leading coefficient of each g_i is 1;*

(ii) *for all $i \neq j$, no $u \in \operatorname{supp}(g_j)$ is divisible by $\mathtt{in}_<(g_i)$.*

Theorem 2.17. *Each ideal I has a unique reduced Gröbner basis.*

Proof. Let $G(\mathtt{in}_<(I)) = \{u_1, \ldots, u_m\}$ and choose $g_1, \ldots, g_m \in I$ with $u_i = \mathtt{in}_<(g_i)$ for $i = 1, \ldots, m$. Then $g_1, \ldots, g_m$ is a Gröbner basis of I, which may, however, not be reduced. Let

$$g_1 = \sum_{i=2}^{m} q_ig_i + h_1 \tag{2.7}$$

be a standard expression of g_1 with respect to $g_2, \ldots, g_m$. Then no $u \in \text{supp}(h_1)$ is divisible by $\texttt{in}_<(g_i)$ for $i = 2, \ldots, m$.

We have $\texttt{in}_<(g_1) \geq \texttt{in}_<(q_i g_i)$ for all i. If we suppose $\texttt{in}_<(g_1) = \texttt{in}_<(q_i g_i)$ for some i, then $u_1 = \texttt{in}_<(q_i g_i) = \texttt{in}_<(q_i)u_i$, a contradiction since $u_1, \ldots, u_m$ is a minimal system of generators of $\texttt{in}_<(I)$. It follows therefore from (2.7) that $\texttt{in}_<(h_1) = \texttt{in}_<(g_1) = u_1$.

Suppose we have already found $h_1, \ldots, h_i \in I$ with the property that $\texttt{in}_<(h_j) = u_j$ for $j = 1, \ldots, i$ and that for all $j \leq i$ no $u \in \text{supp}(h_j)$ is divisible by any u_k with $k \neq j$. If $i < m$, we let h_{i+1} be a remainder of g_{i+1} with respect to $h_1, \ldots, h_i, g_{i+2}, \ldots, g_m$. Then with the same argument as for $i = 1$ we have $\texttt{in}(h_{i+1}) = \texttt{in}_<(g_{i+1}) = u_{i+1}$, and no $u \in \text{supp}(h_{i+1})$ is divisible by any u_k with $k \neq i+1$. Thus step by step we can replace the g_i by the h_i and finally obtain $h_1, \ldots, h_m$ which is again a Gröbner basis of I because $\texttt{in}_<(h_i) = \texttt{in}_<(g_i)$ for $i = 1, \ldots, m$. By construction this new Gröbner basis $h_1, \ldots, h_m$ satisfies condition (ii) in the definition of a reduced Gröbner basis. Let c_i be the leading coefficient of h_i and set $g_i' = c_i^{-1}h_i$. Then obviously $g_1', \ldots, g_m'$ satisfies both conditions (i) and (ii), and hence is a reduced Gröbner basis of I.

Suppose $g_1, \ldots, g_r$ and $g_1', \ldots, g_s'$ are both reduced Gröbner basis of I. Then $G(\texttt{in}_<(I)) = \{\texttt{in}_<(g_1), \ldots, \texttt{in}_<(g_r)\}$, because otherwise condition (ii) would be violated. The same holds true for $g_1', \ldots, g_s'$. It follows that $r = s = m$, and we may assume that $\texttt{in}_<(g_i) = \texttt{in}_<(g_i')$ for $i = 1, \ldots, r$. Assume that $g_j \neq g_j'$ for some j. Then $f = g_j - g_j' \in I$, $f \neq 0$ and $\texttt{in}_<(f) \in \text{supp}(g_j - g_j') \subset \text{supp}(g_j) \cup \text{supp}(g_j')$. Say, $\texttt{in}_<(f) \in \text{supp}(g_j)$. Since $\texttt{in}_<(f) \in \texttt{in}_<(I)$, there exists some i such that $\texttt{in}_<(g_i)$ divides $\texttt{in}_<(f)$. On the other hand, since $\texttt{in}_<(g_j) \notin \text{supp}(g_j - g_j')$, it follows that $\texttt{in}_<(f) < \texttt{in}_<(g_j)$. This implies that $i \neq j$, contradicting condition (ii). □

Problems

In the following problems, K is a field and S denotes the polynomial ring $K[x_1, \ldots, x_n]$.

Problem 2.1. Check that the lexicographic, the pure lexicographic and the reverse lexicographic orders are indeed monomial orders.

Problem 2.2. If we would define the pure reverse lexicographic order in analogy to the pure lexicographic order, would this be a monomial order?

Problem 2.3. Let $\pi : [n] \to [n]$ be the permutation with $\pi(i) = n - i + 1$ for $i = 1, \ldots, n$, and let $<$ be the lexicographic order on S. Show that $<_\pi^{\text{op}}$ is the reverse lexicographic order, where by definition $u <_\pi^{\text{op}} v$, if either $\deg u < \deg v$, or $\deg u = \deg v$ and $v <_\pi u$.

Problem 2.4. Show that there exist infinitely many different monomial orders on a polynomial ring with at least 2 variables.

Problem 2.5. Let $\mathrm{Mon}_d(S)$ denote the set of monomials of S of degree d. A subset $L \subseteq \mathrm{Mon}_d(S)$ is called a **lexsegment**, if for all $u \in L$ and $v > u$ it follows that $v \in L$. Here $>$ is the lexicographic order with the natural order of indeterminates. Give an example of a lexsegment in $L \subseteq \mathrm{Mon}_d(S)$ which is not a lexsegment viewed as a subset of $\mathrm{Mon}_d(S[x_{n+1}])$, and characterize those lexsegments in $\mathrm{Mon}_d(S)$ which are also lexsegments in $\mathrm{Mon}_d(S[x_{n+1}])$.

Problem 2.6. Let $\mathbf{x}^{\mathbf{a}} \in \mathrm{Mon}(S)$ with $\mathbf{x}^{\mathbf{a}} \neq 1$. Describe in terms of $\mathbf{a}$ the largest monomial with respect to the (reverse) lexicographic order which is smaller than $\mathbf{x}^{\mathbf{a}}$.

Problem 2.7. Let $\mathbf{w} = (w_1, \ldots, w_n) \in \mathbb{R}^n_+$ be a weight vector with the property that the entries $w_1, \ldots, w_n$ of $\mathbf{w}$ are linearly independent over $\mathbb{Q}$, and define $\mathbf{x}^{\mathbf{a}} < \mathbf{x}^{\mathbf{b}}$ if and only if $\mathbf{a} \cdot \mathbf{w} < \mathbf{b} \cdot \mathbf{w}$. Show that $<$ is a monomial order.

Problem 2.8. Let $<$ be a monomial order on S, and let $I \subseteq J$ be ideals in S with the property that $\mathtt{in}_<(I) = \mathtt{in}_<(J)$. Show that $I = J$.

Problem 2.9. Let $<$ be a monomial order on S, and let $I \subseteq S$ be a graded ideal in S. We denote by $\mathfrak{m} = (x_1, \ldots, x_n)$ the graded maximal ideal of S. The **Loewy length** of S/I is the infimum of the numbers k such that $\mathfrak{m}^k \subseteq I$. Show that S/I and $S/\mathtt{in}_<(I)$ have the same Loewy length.

Problem 2.10. Let $I \subset S$ be a graded ideal. Show that I admits a Gröbner basis of homogeneous elements.

Problem 2.11. A polynomial $f \in S$ whose support consists of two monomials is called a **binomial** (with coefficients). Let $I \subset S$ be an ideal which is generated by binomials. Show that I admits a Gröbner basis whose elements are all binomials.

Problem 2.12. Compute for the reverse lexicographic order a standard expression of $f = 2x_2^2x_3^3 - x_1x_2^3$ with respect to $g_1 = x_3^2 - 1$, $g_2 = x_2x_3 + x_2^2$ and $g_3 = x_1x_2 - x_3$.

Problem 2.13. Let $g_1, \ldots, g_m$ be a Gröbner basis of $I = (g_1, \ldots, g_m)$. Use Corollary 2.13 to show that $I = S$ if and only if one of the g_i is a nonzero constant polynomial, that is, $g_i = c$ for some $c \in K$, $c \neq 0$.

Problem 2.14. Compute a Gröbner basis of the ideal $I = (x_1^2 - 2x_2x_3, x_1x_2^2 + x_3^2, x_2x_3 + x_3)$ with respect to the lexicographic and the reverse lexicographic order.

Problem 2.15. Let $g_1, \ldots, g_m$ be linear forms in S (i.e. homogeneous polynomials of degree 1), and let $<$ be an arbitrary monomial order on S. Deduce directly from the definition of a Gröbner basis, that if $\mathcal{G} = g_1, \ldots, g_m$ is a minimal system of generators of I, then $\mathcal{G}$ is a Gröbner basis of I if and only if $\mathtt{in}_<(g_i) \neq \mathtt{in}_<(g_j)$ for all $i \neq j$.

Problem 2.16. Let $<$ be a monomial order on S, and let I and J be ideals in S. Then $I = J$ if and only if their reduced Gröbner bases are the same.

Problem 2.17. Let $f, g \in S$ such that $\mathtt{in}_<(f)$ and $\mathtt{in}_<(g)$ are relatively prime and let u and v be any monomials in S. Then $S(uf, vg)$ reduces to zero with respect to uf and ug.

Problem 2.18. Let $I, J \subset S$ be ideals and $<$ a monomial order on S. Let $\mathcal{G}, \mathcal{G}'$ be Gröbner bases of I, respectively J, with respect to $<$. Prove that if $\mathtt{in}_<(g)$ and $\mathtt{in}_<(g')$ are relatively prime for any $g \in \mathcal{G}, g' \in \mathcal{G}'$, then $\mathcal{G} \cup \mathcal{G}'$ is a Gröbner basis of $I + J$.

Chapter 3

First applications

In this chapter we present basic applications of Gröbner bases in polynomial ideal theory. The essential tools are the elimination orders along with the Elimination Theorem.

3.1. Elimination of variables

3.1.1. Elimination orders. Let K be a field and $S = K[x_1, \dots, x_n]$ the polynomial ring in n variables over K.

Definition 3.1. *Let t be an integer, $1 \leq t \leq n$. An* **elimination order** *on S for $x_1, \dots, x_t$ is a monomial order $<$ on S which satisfies the following condition: for any two monomials $u, v \in S$ such that $x_j | u$ for some $1 \leq j \leq t$ and $x_j \not| v$ for all $1 \leq j \leq t$, one has $u > v$.*

In other words, $<$ is an elimination order for the first t variables if any monomial which is divisible by some variable x_j with $1 \leq j \leq t$, is strictly greater than any monomial in the last $n - t$ variables. Obviously, this is equivalent to saying that if a polynomial $f \in S$ has the property that its initial monomial with respect to $<$ belongs to the subring $K[x_{t+1}, \dots, x_n]$ of S, then f is contained in the same subring of S.

The pure lexicographic order is an elimination order for $x_1, \dots, x_t$ for any $1 \leq t \leq n$. A common way to produce elimination orders is to take a product of two arbitrary orders on the subrings $K[x_1, \dots, x_t]$ and $K[x_{t+1}, \dots, x_n]$. Another standard procedure to construct elimination orders comes from Proposition 2.3. Indeed, given a monomial order $<$ on S, we consider the weighted order $<_{\mathbf{w}}$ with respect to the weight vector $\mathbf{w} = (1, \dots, 1, 0, \dots, 0)$ whose first t entries are 1 and the last $n - t$ entries are 0. Proposition 2.3 actually claims that $<_{\mathbf{w}}$ is an elimination order on S.

Definition 3.2. *Let $I \subset S$ be an ideal and $1 \le t \le n$ an integer. The ideal $I_t = I \cap K[x_{t+1}, \ldots, x_n]$ is called the t-th **elimination ideal** of I.*

3.1.2. The Elimination Theorem. The following theorem is the basis for elimination of variables.

Theorem 3.3. *Let $I \subset S$ be an ideal and $1 \le t \le n$ an integer. If $\mathcal{G}$ is a Gröbner basis of I with respect to some elimination order $<$ for $x_1, \ldots, x_t$, then $\mathcal{G}_t = \mathcal{G} \cap K[x_{t+1}, \ldots, x_n]$ is a Gröbner basis of I_t with respect to the induced order on the subring $K[x_{t+1}, \ldots, x_n]$.*

Proof. Let $\mathcal{G} = \{g_1, \ldots, g_s\}$ and assume that $\mathcal{G}_t = \{g_1, \ldots, g_r\}$, that is, $g_1, \ldots, g_r \in K[x_{t+1}, \ldots, x_n]$ and $g_{r+1}, \ldots, g_s \notin K[x_{t+1}, \ldots, x_n]$. In particular, by the choice of the monomial order, we have $\mathtt{in}_<(g_j) \notin K[x_{t+1}, \ldots, x_n]$ for all $j > r$. We show that the set $\{\mathtt{in}_<(g_1), \ldots, \mathtt{in}_<(g_r)\}$ generates $\mathtt{in}_<(I_t)$. Let $f \in I_t$ be a nonzero polynomial. Since $\mathtt{in}_<(f) \in \mathtt{in}_<(I)$ and $\mathtt{in}_<(f)$ is a monomial in the last $n - t$ variables, we must have $\mathtt{in}_<(g_j) | \mathtt{in}_<(f)$ for some $j \le r$, whence $\mathtt{in}_<(f) \in (\mathtt{in}_<(g_1), \ldots, \mathtt{in}_<(g_r))$. Therefore we have $\mathtt{in}_<(I_t) \subset (\mathtt{in}_<(g_1), \ldots, \mathtt{in}_<(g_r))$. The other inclusion is obvious. □

For example, let $I = (x_1^2 - x_2x_3, x_1x_2^2 + x_3^3, x_2x_3 + x_3^2) \subset K[x_1, x_2, x_3]$. The Gröbner basis of I with respect to the pure lexicographic order is $\{x_1^2 - x_2x_3, x_1x_2^2 + x_3^3, x_1x_3^3 + x_2^3x_3, x_2x_3 + x_3^2, x_3^4\}$. It follows that $I \cap K[x_2, x_3] = (x_2x_3 + x_3^2, x_3^4)$ and $I \cap K[x_3] = (x_3^4)$.

Here is a reformulation of the Elimination Theorem in terms of initial ideals.

Corollary 3.4. *Let $I \subset S$ be an ideal and $1 \le t \le n$ an integer. Then*

$$\mathtt{in}_<(I \cap K[x_{t+1}, \ldots, x_n]) = \mathtt{in}_<(I) \cap K[x_{t+1}, \ldots, x_n]$$

for any elimination order $<$ on S for $x_1, \ldots, x_t$.

3.2. Applications to operations on ideals

In this section we use Theorem 3.3 to get algorithms for operations with polynomial ideals.

3.2.1. Intersection of ideals. In order to get a procedure to compute a Gröbner basis for the intersection of two polynomial ideals we first need to introduce a notation. For an ideal I of S and a polynomial $p(t)$ in a new variable t, we denote by $p(t)I$ the ideal of $S[t]$ which is generated by $\{p(t)f : f \in I\} \subset S[t]$. It is straightforward to notice that if I is generated by $f_1, \ldots, f_r$ as an ideal of S, then $p(t)I$ is generated by $p(t)f_1, \ldots, p(t)f_r$ as an ideal in $S[t]$.

Proposition 3.5. *Let $I, J \subset S$ be ideals. Then $I \cap J = (tI + (1-t)J) \cap S$.*

Proof. Let $f \in I \cap J$. Then $f = tf + (1-t)f \in (tI + (1-t)J) \cap S$. For the other inclusion, let $g \in (tI+(1-t)J)\cap S$. We write $g = h_1 + h_2$, where $h_1 \in tI$ and $h_2 \in (1-t)J$. Hence there exist some polynomials $h_1' \in IS[t], h_2' \in JS[t]$ such that $h_1 = th_1'$ and $h_2 = (1-t)h_2'$. It follows that $g = th_1' + (1-t)h_2'$. Substituting t by zero in this equality we get $g = h_2'(0, x_1, \ldots, x_n) \in J$. Next, substituting t by 1, we get $g = h_1'(1, x_1, \ldots, x_n) \in I$. Consequently, we have $g \in I \cap J$. □

The above proposition provides a procedure to compute a Gröbner basis for $I \cap J$ if some generating systems for I and J are given. Indeed, if I is generated by some polynomials $f_1, \ldots, f_r$ and J is generated by $g_1, \ldots, g_s$, then $tf_1, \ldots, tf_r, (1-t)g_1, \ldots, (1-t)g_s$ generate $tI + (1-t)J$. We choose an elimination order for the variable t on $S[t]$ and compute a Gröbner basis $\mathcal{G}$ of $tI + (1-t)J$. By Theorem 3.3, it follows that $\mathcal{G} \cap S$ is a Gröbner basis of $I \cap J$.

For example, let $I = (x_1x_3 - x_2^2, x_1^3 - x_2x_3)$, $J = (x_1, x_3^2) \subset K[x_1, x_2, x_3]$. The reduced Gröbner basis of $tI + (1-t)J \subset K[t, x_1, x_2, x_3]$ with respect to the pure lexicographic order with $t > x_1 > x_2 > x_3$ is $\{tx_1 - x_1, tx_2^2 - x_1x_3, tx_2x_3 - x_1^3, tx_3^2 - x_3^2, x_1^4 - x_1x_2x_3, x_1^3x_2 - x_1x_3^2, x_1^2x_2^2 - x_2x_3^2, x_1^2x_3 - x_1x_2^2, x_1x_2^4 - x_2x_3^3, x_1x_3^3 - x_2^2x_3^2, x_2^6x_3^2 - x_2x_3^6\}$.

Eliminating the variable t we get $I \cap J = (x_1^4 - x_1x_2x_3, x_1^3x_2 - x_1x_3^2, x_1^2x_2^2 - x_2x_3^2, x_1^2x_3 - x_1x_2^2, x_1x_2^4 - x_2x_3^3, x_1x_3^3 - x_2^2x_3^2, x_2^6x_3^2 - x_2x_3^6)$.

3.2.2. Ideal quotient. Let $I, J \subset S$ be two polynomial ideals with $J = (g_1, \ldots, g_s)$. In order to determine a Gröbner basis of $I \colon J$, one first uses the equality $I \colon J = \bigcap_{i=1}^s I \colon (g_i)$. Therefore we can reduce the problem to the case when J is a principal ideal.

Proposition 3.6. *Let $I \subset S$ be an ideal and $g \in S$ a polynomial. If $\{q_1, \ldots, q_r\}$ is a set of generators for $I \cap (g)$, then $\{q_1/g, \ldots, q_r/g\}$ is a set of generators for $I \colon (g)$.*

Proof. Let $\mathcal{H} = \{q_1/g, \ldots, q_r/g\}$. It is obvious that $(g)\mathcal{H} \subset (q_1, \ldots, q_r) \subset I$, hence $(\mathcal{H}) \subset I \colon (g)$. For the reverse inclusion, let $f \in I \colon (g)$. It follows that $fg \in I \cap (g)$, thus $fg \in (q_1, \ldots, q_r)$. As all the polynomials $q_1, \ldots, q_r$ are divisible by g, it follows that $f \in (\mathcal{H})$. □

We illustrate the above procedure in the following example. Let $I = (x_1x_3 - x_2^2, x_1^3 - x_2x_3)$, $J = (x_1, x_3) \subset K[x_1, x_2, x_3]$. We consider the pure lexicographic order. We have

$$I \colon J = (I \colon (x_1)) \cap (I \colon (x_3)).$$

Applying the procedure given by the above proposition and the algorithm for computing the intersection we get

$$I\colon (x_1) = (x_1^3 - x_2x_3, x_1^2x_2 - x_3^2, x_1x_2^3 - x_3^3, x_1x_3 - x_2^2, x_2^5 - x_3^4)$$

and

$$I\colon (x_3) = (x_1^3 - x_2x_3, x_1^2x_2^2 - x_2x_3^2, x_1x_2^4 - x_2x_3^3, x_1x_3 - x_2^2, x_2^6 - x_3^5).$$

The reduced Gröbner basis of $I\colon J$ with respect to the pure lexicographic order is the set

$$\{x_1^3 - x_2x_3, x_1^2x_2^2 - x_2x_3^2, x_1x_2^4 - x_2x_3^3, x_1x_3 - x_2^2, x_2^6 - x_2x_3^4\}.$$

3.2.3. Saturation and radical membership. For a graded ideal I of S, in Subsection 1.2.1, we defined its saturation by $I^{\text{sat}} = I\colon \mathfrak{m}^\infty = \bigcup_{i=1}^{\infty} I\colon \mathfrak{m}^i$. Of course one can compute a Gröbner basis of I^{sat} by applying the procedures from Subsection 3.2.1 and Subsection 3.2.2 to $I\colon \mathfrak{m}^i$ and getting successively the reduced Gröbner bases of $I\colon \mathfrak{m}^i$ for $i \geq 1$. When this sequence of ideals stabilizes, one gets the Gröbner basis of the saturation. We describe here another way to compute the saturation. For an ideal I of S and a polynomial $f \in S$, we consider

$$I\colon f^\infty = \{g \in S : \text{ there exists } \ i > 0 \text{ such that } f^i g \in I\}.$$

One may easily check that $I\colon f^\infty$ is an ideal of S which is homogeneous if I and f are homogeneous, and $I\colon f^\infty = \bigcup_{i=1}^{\infty} I\colon f^i$. We call this ideal the **saturation of I with respect to** f. By Problem 1.9 we have $I^{\text{sat}} = \bigcap_{i=1}^{n}(I\colon x_i^\infty)$. Hence it is enough to give a procedure for computing a Gröbner basis of the saturation of an ideal I with respect to a polynomial f.

Proposition 3.7. *Let $I \subset S$ be an ideal and $f \in S$ a polynomial. Let t be a new variable and $\widetilde{I}$ the ideal generated in $S[t]$ by I and the polynomial $1 - ft$. Then $I\colon f^\infty = \widetilde{I} \cap S$.*

Proof. Let $g \in I\colon f^\infty$. There exists $i \geq 1$ such that $u = gf^i \in I$. Then

$$g = gf^it^i + (1 - f^it^i)g = ut^i + (1 - ft)(1 + ft + \cdots + f^{i-1}t^{i-1})g \in (I, 1 - ft).$$

This shows that $I\colon f^\infty \subset \widetilde{I} \cap S$.

For the opposite inclusion, let $g \in \widetilde{I} \cap S$. We can write $g = u + v(1 - tf)$, where $u \in IS[t]$ and $v \in S[t]$. If I is generated by $f_1, \ldots, f_s$, then $u = a_1f_1 + \cdots + a_sf_s$ for some polynomials $a_1, \ldots, a_s \in S[t]$. Hence

$$g = a_1f_1 + \cdots + a_sf_s + v(1 - tf).$$

Substituting t by $1/f$ in the above expression of g, we obtain

$$g = a_1(\frac{1}{f}, x_1, \ldots, x_n)f_1 + \cdots + a_s(\frac{1}{f}, x_1, \ldots, x_n)f_s.$$

Clearing the denominators in the above equality, we find an enough large i such that $f^i g$ can be expressed as a combination of $f_1, \ldots, f_s$ with coefficients in S, which implies that $f^i g \in I$, that is, $g \in I\colon f^\infty$. □

Therefore, to find a Gröbner basis of $I\colon f^\infty$, we need to eliminate the variable t from the Gröbner basis of the ideal $(I, 1-ft) \subset S[t]$, thus we need to consider an elimination order for t.

Proposition 3.7 can be used to test **radical membership**. Indeed, it is obvious that $f \in \sqrt{I}$ if and only if $I\colon f^\infty = S$, that is, if the reduced Gröbner basis of $\widetilde{I}$ is $\{1\}$.

3.2.4. K-algebra homomorphisms. Let $S = K[x_1, \ldots, x_n]$ and $S' = K[y_1, \ldots, y_m]$ be two polynomial algebras, $f_1, \ldots, f_n$ polynomials in S', and $I' \subset S'$ an ideal. By Theorem 1.1, we have a K-algebra homomorphism $\psi : S \longrightarrow S'/I'$ with $\psi(x_i) = f_i + I'$ for all $1 \le i \le n$. Then, for any ideal $I \subset \operatorname{Ker}(\psi)$, there exists a K-algebra homomorphism $\varphi : S/I \longrightarrow S'/I'$ with the property that $\varphi(x_i + I) = \psi(x_i) = f_i + I'$ for all i.

In the first step we are interested in finding a generating set for the kernel of φ by using elimination.

Proposition 3.8. *Let $\varphi : S/I \longrightarrow S'/I'$ be a K-algebra homomorphism with $\varphi(x_i + I) = f_i + I'$ for $1 \le i \le n$, where $f_1, \ldots, f_n \in S'$. Let $R = K[x_1, \ldots, x_n, y_1, \ldots, y_m]$, $J = I'R + (x_1 - f_1, \ldots, x_n - f_n)$, and $L = J \cap S$. Then $\operatorname{Ker}(\varphi)$ is the image of the ideal L in S/I, that is $\operatorname{Ker}(\varphi) = (I + L)/I$.*

Proof. Let $g + I \in \operatorname{Ker}(\varphi)$, that is, $\varphi(g + I) = 0$. It follows that $g(f_1, \ldots, f_n) + I' = 0$, hence $g(f_1, \ldots, f_n) \in I'$. If $g = \sum c_{\mathbf{a}} \mathbf{x}^{\mathbf{a}}$, then we can write

$$g = \sum c_{\mathbf{a}} (x_1 - f_1 + f_1)^{a_1} \cdots (x_n - f_n + f_n)^{a_n} = g' + \sum c_{\mathbf{a}} \mathbf{f}^{\mathbf{a}} = g' + g(f_1, \ldots, f_n),$$

with $g' \in (x_1 - f_1, \ldots, x_n - f_n)$. Therefore, $g \in J \cap S = L$ and $g + I \in (I + L)/I$. Conversely, let $g \in L = J \cap S$ and $\{h_1, \ldots, h_r\} \subset S'$ be a set of generators of I'. Then there exist some polynomials $a_1, \ldots, a_r, b_1, \ldots, b_n \in R$ such that

$$g = a_1 h_1 + \cdots + a_r h_r + b_1(x_1 - f_1) + \cdots + b_n(x_n - f_n).$$

Substituting x_i by f_i in the above equality for $1 \le i \le n$, we get

$$g(f_1, \ldots, f_n) \in (h_1, \ldots, h_r) = I'.$$

Therefore $\varphi(g + I) = g(f_1, \ldots, f_n) + I' = 0$, whence $g + I \in \operatorname{Ker}(\varphi)$. □

Corollary 3.9. *Let* $f_1, \dots, f_n \in S'$ *be polynomials and* $\varphi : S \longrightarrow S'$ *the* K*-algebra homomorphism defined by* $x_i \mapsto f_i$ *for* $1 \leq i \leq n$. *Let* $J = (x_1 - f_1, \dots, x_n - f_n) \subset R = K[x_1, \dots, x_n, y_1, \dots, y_m]$. *Then* $\mathrm{Ker}(\varphi) = J \cap S$.

Proof. Take $I = I' = (0)$ in the above proposition. □

Corollary 3.9 solves the following problem. Given a family of parametric equations $x_1 = f_1, \dots, x_n = f_n$ with $f_1, \dots, f_n \in S'$, we may find the implicit relations among the polynomials $f_1, \dots, f_n$.

For example, let $\varphi : K[x_1, x_2] \longrightarrow K[y]$ be defined by $\varphi(x_1) = y^2, \varphi(x_2) = y^3$. The reduced Gröbner basis of $J = (x_1 - y^2, x_2 - y^3) \subset K[y, x_1, x_2]$ with respect to the pure lexicographic order with $y > x_1 > x_2$ is $\{y^2 - x_1, yx_1 - x_2, yx_2 - x_1^2, x_1^3 - x_2^2\}$, therefore $\mathrm{Ker}(\varphi) = (x_1^3 - x_2^2)$.

Example 3.10. Let $I \subset S$ be a graded ideal generated by some homogeneous polynomials $f_1, \dots, f_q$ and t a new indeterminate over K. The **Rees ring** of I, denoted by $\mathcal{R}(I)$, is the graded subring of $S[t]$ given by

$$\mathcal{R}(I) = \bigoplus_{j \geq 0} I^j t^j = S[f_1 t, \dots, f_q t].$$

Consider the presentation of $\mathcal{R}(I)$,

$$\varphi : R = S[u_1, \dots, u_q] \longrightarrow \mathcal{R}(I),$$

defined by

$$x_i \mapsto x_i \text{ for } 1 \leq i \leq n \text{ and } u_j \mapsto f_j t \text{ for } 1 \leq j \leq q.$$

$J = \mathrm{Ker}(\varphi)$ is an ideal of $S[u_1, \dots, u_q]$ which is called **the presentation ideal** of $\mathcal{R}(I)$. By Corollary 3.9, we have

$$J = (u_1 - f_1 t, \dots, u_q - f_q t) \cap R.$$

Let $\mathcal{G}$ be a Gröbner basis of $(u_1 - f_1 t, \dots, u_q - f_q t)$ with respect to an elimination order for t in the polynomial ring $K[t, x_1, \dots, x_n, u_1, \dots, u_q]$. Then $\mathcal{G} \cap R$ is a Gröbner basis of the presentation ideal J.

In the following proposition we study the image of a K-algebra homomorphism.

Proposition 3.11. *Let* $\varphi : S/I \longrightarrow S'/I'$ *be a* K*-algebra homomorphism with* $\varphi(x_i + I) = f_i + I'$ *for* $1 \leq i \leq n$, *where* $f_1, \dots, f_n \in S'$. *Let* $R = K[x_1, \dots, x_n, y_1, \dots, y_m]$ *be endowed with an elimination order* $<$ *for the indeterminates* $y_1, \dots, y_m$ *and* $J = I'R + (x_1 - f_1, \dots, x_n - f_n) \subset R$. *Let* $\mathcal{G} = \{g_1, \dots, g_r\}$ *be a reduced Gröbner basis of* J *with respect to* $<$. *Let* $g \in S'$ *be a polynomial and* h *its remainder with respect to* $\mathcal{G}$. *Then:*

(i) $g + I' \in \mathrm{Im}(\varphi)$ *if and only if* $h \in S$.

(ii) *If* $g + I' \in \mathrm{Im}(\varphi)$, *then* $g + I' = \varphi(h + I)$.

Proof. Let $g = a_1g_1 + \cdots + a_rg_r + h$ be a standard expression of g with $h \in S$, where $a_1, \dots, a_r \in R$. Substituting x_i by f_i in the above expression, we get $g - h(f_1, \dots, f_n) \in I'$ since $g_i(f_1, \dots, f_n, y_1, \dots, y_m) \in I'$ for all $1 \leq i \leq r$. Therefore $g + I' = h(f_1, \dots, f_n) + I' = \varphi(h + I)$. Hence we proved (ii) and the "if" part of (i).

For the "only if" part of (i), let $g \in S'$ such that $g + I' \in \text{Im}(\varphi)$. Thus there exists $q \in S$ such that $g + I' = \varphi(q + I) = q(f_1, \dots, f_n) + I'$. It follows that $g - q(f_1, \dots, f_n) \in I'R \subset J$. Therefore, as polynomials in R, g and $q(f_1, \dots, f_n)$ have the same remainder with respect to $\mathcal{G}$. On the other hand, $q - q(f_1, \dots, f_n) \in (x_1 - f_1, \dots, x_n - f_n) \subset J$. Thus q and $q(f_1, \dots, f_n)$ have the same remainder with respect to $\mathcal{G}$. It follows that q has the remainder h with respect to $\mathcal{G}$. Since $<$ is an elimination order for $y_1, \dots, y_m$ and q is a polynomial in $x_1, \dots, x_n$, it follows that $h \in S$. Indeed, let $\mathcal{G} = \{g_1, \dots, g_s\}$ and let

$$q = a_1g_1 + \cdots + a_sg_s + h \tag{3.1}$$

be a standard expression of q with respect to $\mathcal{G}$. Using condition (ii) from Theorem 2.11 we have $\mathtt{in}_<(a_ig_i) \leq \mathtt{in}_<(q)$ for all $1 \leq i \leq s$. It follows that $\mathtt{in}_<(a_ig_i)$ are monomials in $x_1, \dots, x_n$, and, consequently, a_ig_i are polynomials in S as well for all i. From equality (3.1) it follows that $h \in S$. □

As a consequence of the above proposition one can provide a **subalgebra membership** test. Let $f_1, \dots, f_n$ be polynomials in S'. Given a polynomial $g \in S'$, we would like to decide whether g belongs or does not to the subalgebra $K[f_1, \dots, f_n]$ of S'. The answer to this problem is given in the following

Corollary 3.12. *Let $f_1, \dots, f_n$ be polynomials in $S' = k[y_1, \dots, y_m]$. Let $J = (x_1 - f_1, \dots, x_n - f_n) \subset R = K[x_1, \dots, x_n, y_1, \dots, y_m]$ and $\mathcal{G}$ a Gröbner basis of J with respect to some elimination order for $y_1, \dots, y_m$. Let g be a polynomial in S' and h its remainder with respect to $\mathcal{G}$. Then:*

(i) *$g \in K[f_1, \dots, f_n]$ if and only if $h \in S$.*

(ii) *If $g \in K[f_1, \dots, f_n]$, then $g = h(f_1, \dots, f_n)$ is a representation of g as a polynomial in $f_1, \dots, f_n$.*

Proof. Take $I = I' = (0)$ in Proposition 3.11. □

Let $s_1, \dots, s_n \in S$ be the elementary symmetric polynomials in $x_1, \dots, x_n$. We recall that, for $1 \leq k \leq n$, $s_k = \sum_{1 \leq i_1 < \cdots < i_k \leq n} x_{i_1} \cdots x_{i_k}$. By the Fundamental Theorem of Symmetric Polynomials, we know that $f \in S$ is symmetric if and only if $f \in K[s_1, \dots, s_n]$. Corollary 3.12 provides a procedure for expressing a symmetric polynomial as a polynomial in $s_1, \dots, s_n$.

Let us take, for example, $f = x_1^4 + x_2^4 + x_3^4 - x_1^2x_2^2 - x_1^2x_3^2 - x_2^2x_3^2 \in \mathbb{Q}[x_1, x_2, x_3]$. f is obviously a symmetric polynomial. We consider the monomial ordering on $\mathbb{Q}[x_1, x_2, x_3, t_1, t_2, t_3]$ given by the product of the reverse lexicographic orders on $\mathbb{Q}[x_1, x_2, x_3]$ and $\mathbb{Q}[t_1, t_2, t_3]$. This is an elimination order for x_1, x_2, x_3. With respect to this order, the ideal $J = (t_1 - s_1, t_2 - s_2, t_3 - s_3)$ of $\mathbb{Q}[x_1, x_2, x_3, t_1, t_2, t_3]$ has the following reduced Gröbner basis

$$\mathcal{G} = \{x_1 + x_2 + x_3 - t_1, x_2^2 - x_1x_3 - x_2t_1 + t_2, x_3^3 - x_3^2t_1 + x_3t_2 - t_3\}$$

and the remainder of f with respect to $\mathcal{G}$ is

$$h = t_1^4 - 4t_1^2t_2 + t_2^2 + 6t_1t_3,$$

which gives the expression of f as a polynomial in s_1, s_2, s_3, namely

$$f = s_1^4 - 4s_1^2s_2 + s_2^2 + 6s_1s_3.$$

3.2.5. Homogenization. Let $f \in S$ be a polynomial of degree d and $f = \sum_{i=0}^d f_i$ its decomposition into homogeneous components. Then

$$f^h = \sum_{i=0}^{d} t^{d-i} f_i$$

is a homogeneous polynomial of degree d in the ring $S[t]$. f^h is called the **homogenization of** f. By Problem 3.12, f^h can be computed by using the formula

$$f^h = t^d f(\frac{x_1}{t}, \dots, \frac{x_n}{t}).$$

For example, let $f = 1 - x_2 + x_1^2 - x_1x_2 + x_1^2x_3 \in K[x_1, x_2, x_3]$. Its homogenization is $f^h = t^3 - t^2x_2 + tx_1^2 - tx_1x_2 + x_1^2x_3$.

If $g \in S[t]$ is a homogeneous polynomial, we denote by $\bar{g}$ its **dehomogenization**, that is, the polynomial of S which is obtained from g by the substitution $t \mapsto 1$. It is obvious that for any polynomial $f \in S$ we have $\bar{f^h} = f$ and for any homogeneous polynomial $g \in S[t]$, $g = t^m \bar{g}^h$ for some $m \geq 0$.

Definition 3.13. *Let $I \subset S$ be an ideal. The* **homogenization** *of I is the ideal I^h generated by $\{f^h : f \in I\}$.*

Note that if I is generated by $f_1, \dots, f_s$, then I^h is not necessarily generated by $f_1^h, \dots, f_s^h$.

For example, let $I = (f_1, f_2) \subset K[x_1, x_2, x_3]$, where $f_1 = x_2 - x_1^2$ and $f_2 = x_3 - x_1^3$ (the twisted cubic). We have $f_1^h = tx_2 - x_1^2$ and $f_2^h = t^2x_3 - x_1^3$. On the other hand, $f = f_2 - x_1f_1 = x_3 - x_1x_2 \in I$ has the homogenization $f^h = tx_3 - x_1x_2 \notin (f_1^h, f_2^h)$.

The next lemma characterizes the homogeneous polynomials in $S[t]$ which belong to I^h.

Lemma 3.14. *Let $I \subset S$ be an ideal and $f \in S[t]$ a homogeneous polynomial. Then $f \in I^h$ if and only if $f = t^m g^h$ for some $g \in I$ and some $m > 0$.*

Proof. Let $f \in I^h$. Then $f = \sum_{i=1}^r g_i f_i^h$ with $f_i \in I$ and $g_i \in S[t]$ homogeneous. One gets the following equality by dehomogenization

$$\bar{f} = \sum_{i=1}^{r} \bar{g}_i \bar{f_i^h} = \sum_{i=1}^{n} \bar{g}_i f_i \in I.$$

Since $f = t^m \bar{f}^h$ for some $m \in \mathbb{Z}_+$, we may take $g = \bar{f}$.

The other implication is obvious. □

A monomial order $<$ on S is called **graded** if, for any monomials $u, v \in S$, $\deg u < \deg v$ implies $u < v$. The lexicographic and reverse lexicographic orders are examples of graded monomial orders. Any graded monomial order on S can be extended to $S[t]$ by taking the product order on S together with the natural order of the powers of the variable t, namely:

$$\mathbf{x}^{\mathbf{a}} t^c <' \mathbf{x}^{\mathbf{b}} t^d \text{ if and only if (i) } \mathbf{x}^{\mathbf{a}} < \mathbf{x}^{\mathbf{b}}, \text{ or (ii) } \mathbf{x}^{\mathbf{a}} = \mathbf{x}^{\mathbf{b}} \text{ and } c < d.$$

By the definition of homogenization it follows that $\mathtt{in}_<(g) = \mathtt{in}_{<'}(g^h)$ for all nonzero polynomials $g \in S$.

Proposition 3.15. *Let $I \subset S$ be an ideal and $\mathcal{G} = \{g_1, \ldots, g_s\}$ a Gröbner basis of I with respect to a graded monomial order $<$ on S. Then $\mathcal{G}^h = \{g_1^h, \ldots, g_s^h\}$ is a Gröbner basis of I^h with respect to $<'$.*

Proof. Since I^h is a homogeneous ideal, it suffices to show that for any homogeneous polynomial $f \in I^h$ we have $\mathtt{in}_{<'}(f) \in (\mathtt{in}_{<'}(g_1^h), \ldots, \mathtt{in}_{<'}(g_s^h))$. Indeed, for an arbitrary polynomial $f \in I^h$, we have $\mathtt{in}_{<'}(f) = \mathtt{in}_{<'}(f_j)$ for some homogeneous component f_j of f. Since I^h is a graded ideal, it follows from Proposition 1.2 that $f_j \in I^h$.

By Lemma 3.14, we have $f = t^m g^h$ for some $g \in I$ and some $m \in \mathbb{Z}_+$. Therefore,

$$\mathtt{in}_{<'}(f) = t^m \,\mathtt{in}_{<'}(g^h) = t^m \,\mathtt{in}_<(g).$$

Since $\mathcal{G}$ is a Gröbner basis of I with respect to $<$, there exists a monomial u such that $\mathtt{in}_<(g) = u \,\mathtt{in}_<(g_i)$ for some i, and since $\mathtt{in}_<(g_i) = \mathtt{in}_{<'}(g_i^h)$, we obtain

$$\mathtt{in}_{<'}(f) = t^m u \,\mathtt{in}_{<'}(g_i^h),$$

which ends the proof. □

For example, let $I = (x_2 - x_1^2, x_3 - x_1^3)$ be the ideal of the twisted cubic. The reduced Gröbner basis of I with respect to the reverse lexicographic order is $\{x_1^2 - x_2, x_1x_2 - x_3, x_2^2 - x_1x_3\}$. Thus the Gröbner basis of I^h with respect to the induced order on $S[t]$ is $\{x_1^2 - tx_2, x_1x_2 - tx_3, x_2^2 - x_1x_3\}$.

3.3. Zero dimensional ideals

Throughout this section the field K will be always algebraically closed. We recall that a field K is algebraically closed if any nonconstant polynomial $f \in K[x]$ has a root in K. It turns out that this is equivalent to saying that any nonconstant polynomial $f \in K[x]$ has all its roots in K. The Fundamental Theorem of Algebra states that the field $\mathbb{C}$ of complex numbers is algebraically closed.

Let $I \subset S = K[x_1, \ldots, x_n]$ be an ideal. The set

$$\mathcal{V}(I) = \{a = (a_1, \ldots, a_n) \in K^n : f(a) = 0 \text{ for all } f \in I\}$$

is called an **affine algebraic variety** in K^n. It is easily seen that $\mathcal{V}(I) = \mathcal{V}(f_1, \ldots, f_r)$ for any set of generators $\{f_1, \ldots, f_r\}$ of I, where

$$\mathcal{V}(f_1, \ldots, f_r) = \{a = (a_1, \ldots, a_n) \in K^n : f_i(a) = 0 \text{ for all } 1 \leq i \leq r\}.$$

Let us first consider ideals in polynomial rings of one variable. For an arbitrary ideal $I \subset K[x]$ there exists a polynomial $f \in K[x]$ such that $I = (f)$, hence $\mathcal{V}(I) = \mathcal{V}(f)$. Over an algebraically closed field, we have $\mathcal{V}(f) = \emptyset$ if and only if $f \in K - \{0\}$, that is, if and only if $I = (1)$.

One may ask if the above equivalence, namely $\mathcal{V}(I) = \emptyset \Leftrightarrow I = (1)$, is still true in the ring of polynomials in many variables. The celebrated Hilbert's Nullstellensatz gives a positive answer to this question.

Theorem 3.16 (Weak Hilbert's Nullstellensatz). *Let $I \subset S$ be an ideal. Then $\mathcal{V}(I) = \emptyset$ if and only if $I = (1)$.*

For the proof we refer the reader to [**M86**].

The correspondence $I \mapsto \mathcal{V}(I)$ defines an inclusion reversing map from the set of ideals of S to the set of affine varieties of K^n. One may define as well a correspondence between the subsets of K^n and the set of ideals of S as follows:

$$K^n \supset Z \mapsto \mathcal{I}(Z) \subset S,$$

where $\mathcal{I}(Z) = \{f \in S : f(a) = 0 \text{ for all } a \in Z\}$. One can easily check that $\mathcal{I}(Z)$ is an ideal of S and that $Z_1 \subset Z_2$ implies $\mathcal{I}(Z_1) \supset \mathcal{I}(Z_2)$.

Example 3.17. Let $\mathbf{a} = (a_1, \ldots, a_n) \in K^n$ and $Z = \{\mathbf{a}\}$. Then we have $\mathcal{I}(Z) = (x_1 - a_1, \ldots, x_n - a_n)$. Indeed, obviously, all the generators $x_i - a_i$

vanish at $\mathbf{a}$. Conversely, let $f \in \mathcal{I}(Z)$, that is, $f(\mathbf{a}) = 0$. Then, we have $f = f - f(\mathbf{a}) \in (x_1 - a_1, \dots, x_n - a_n)$.

For $\mathbf{a} = (a_1, \dots, a_n) \in K^n$ we denote by $\mathfrak{m}_\mathbf{a}$ the ideal $(x_1 - a_1, \dots, x_n - a_n)$. $\mathfrak{m}_\mathbf{a}$ is a maximal ideal of S since we have a canonical K-algebra isomorphism $S/\mathfrak{m}_\mathbf{a} \cong K$ induced by the surjective substitution morphism $x_i \mapsto a_i$ for $1 \leq i \leq n$.

The correspondence between ideals and affine varieties is given by the strong version of the Hilbert's Nullstellensatz which follows from the weak version.

Theorem 3.18 (Strong Hilbert's Nullstellensatz). *Let $I \subset S$ be an ideal. Then $\mathcal{I}(\mathcal{V}(I)) = \sqrt{I}$, that is, for any polynomial $f \in S$, $f \in \mathcal{I}(\mathcal{V}(I))$ if and only if there exists an integer $m > 0$ such that $f^m \in I$.*

Proof. Let $m > 0$ such that $f^m \in I$. Then $f^m(\mathbf{a}) = 0$ for all $\mathbf{a} \in \mathcal{V}(I)$, whence $f(\mathbf{a}) = 0$ for all $\mathbf{a} \in \mathcal{V}(I)$. Therefore $f \in \mathcal{I}(\mathcal{V}(I))$. Conversely, let $f \in \mathcal{I}(\mathcal{V}(I))$ and assume that I is generated by $f_1, \dots, f_s$. Then $f(\mathbf{a}) = 0$ for all the common roots $\mathbf{a}$ of $f_1, \dots, f_s$. Let us consider the ideal

$$\widetilde{I} = (f_1, \dots, f_s, 1 - yf) \subset S[y].$$

We show that $\mathcal{V}(\widetilde{I}) = \emptyset$. Indeed, let $(a_1, \dots, a_n, b) \in K^{n+1}$.

If $(a_1, \dots, a_n) \in \mathcal{V}(I)$, then $f(a_1, \dots, a_n) = 0$ since $f \in \mathcal{I}(\mathcal{V}(I))$, thus $(1 - yf)(a_1, \dots, a_n, b) \neq 0$, that is, $(a_1, \dots, a_n, b) \notin \mathcal{V}(\widetilde{I})$.

If $(a_1, \dots, a_n) \notin \mathcal{V}(I)$, then it is obvious that $(a_1, \dots, a_n, b) \notin \mathcal{V}(\widetilde{I})$. Therefore $\mathcal{V}(\widetilde{I}) = \emptyset$.

By the Weak Nullstellensatz, it follows that $\widetilde{I} = (1)$, thus there exist some polynomials $p_1, \dots, p_s, q \in S[y]$ such that

$$1 = p_1 f_1 + \cdots + p_s f_s + q(1 - yf).$$

Substituting $y \mapsto \frac{1}{f}$ in the above equality we get the following identity:

$$1 = p_1(x_1, \dots, x_n, \frac{1}{f}) f_1 + \cdots + p_s(x_1, \dots, x_n, \frac{1}{f}) f_s.$$

Multiplying the above equality by a power large enough of f we obtain an equality of the form

$$f^m = \sum_{i=1}^{s} g_i f_i,$$

where $g_1, \dots, g_s \in S$, thus $f^m \in I$, that is, $f \in \sqrt{I}$. □

The following proposition shows that, over an algebraically closed field, the maximal ideals of the polynomial ring S are in one-to-one correspondence with the points of the affine space K^n.

Proposition 3.19. *Any maximal ideal of S is of the form*

$$\mathfrak{m}_{\mathbf{a}} = (x_1 - a_1, \ldots, x_n - a_n)$$

for some $\mathbf{a} = (a_1, \ldots, a_n) \in K^n$.

Proof. Let $\mathfrak{m} \subset S$ be a maximal ideal. As $\mathfrak{m} \neq (1)$, by Theorem 3.16, we have $\mathcal{V}(\mathfrak{m}) \neq \emptyset$. Let $\mathbf{a} \in \mathcal{V}(\mathfrak{m})$. Then $\mathfrak{m}_{\mathbf{a}} = \mathcal{I}(\{\mathbf{a}\}) \supset \mathcal{I}(\mathcal{V}(\mathfrak{m})) = \sqrt{\mathfrak{m}} \supset \mathfrak{m}$. Since $\mathfrak{m}_{\mathbf{a}}$ and $\mathfrak{m}$ are maximal ideals, it follows that $\mathfrak{m} = \mathfrak{m}_{\mathbf{a}}$. □

Remark 3.20. Let $I \subset S$ be an ideal. Then it is clear that $a \in \mathcal{V}(I)$ if and only if $\mathfrak{m}_a \supset I$, hence the points of the affine variety $\mathcal{V}(I)$ are in one-to-one correspondence with the maximal ideals of S which contain I.

Definition 3.21. *An ideal I in S is called a* ***zero-dimensional ideal*** *if $\mathcal{V}(I)$ is a finite set.*

In the sequel we characterize the zero-dimensional ideals of S in terms of Gröbner bases.

Theorem 3.22. *Let $I \subset S$ be an ideal. The following statements are equivalent:*

(a) *I is a zero-dimensional ideal, that is, $\mathcal{V}(I)$ is finite.*

(b) *I is contained in finitely many maximal ideals of S.*

(c) *S/I is a K-vector space of finite dimension.*

(d) *For any monomial order $<$ on S, the set* $\mathrm{Mon}(S) \setminus \mathrm{Mon}(\mathtt{in}_<(I))$ *is a finite set.*

(e) *If $\mathcal{G}$ is a Gröbner basis with respect to some monomial order $<$ on S, then, for any $1 \leq i \leq n$, there exists $g \in \mathcal{G}$ with $\mathtt{in}_<(g) = x_i^{\nu_i}$ for some $\nu_i \geq 0$.*

Proof. (a) and (b) are equivalent by Remark 3.20.

(c) and (d) are equivalent by Macaulay's Theorem; see Theorem 2.6.

(e) $\Rightarrow$ (d) is obvious since there are only finitely many monomials $w = x_1^{b_1} \cdots x_n^{b_n} \in S$ such that $b_i < \nu_i$ for all $1 \leq i \leq n$.

For (d) $\Rightarrow$ (e), let us fix $1 \leq i \leq n$. As the set $\mathrm{Mon}(S) \setminus \mathrm{Mon}(\mathtt{in}_<(I))$ is finite, there exists some $\nu_i \geq 0$ such that $x_i^{\nu_i}$ is a minimal generator of $\mathtt{in}_<(I)$, whence (e) follows.

Let us prove now the implication (a) $\Rightarrow$ (e). If $\mathcal{V}(I) = \emptyset$, then $I = (1)$ by the Weak Hilbert's Nullstellensatz, and (e) is obviously true. Let $\mathcal{V}(I) \neq \emptyset$ and assume that $\mathcal{V}(I) = \{a_1, \ldots, a_t\}$ for some $t \geq 1$, where $a_i = (a_{i1}, \ldots, a_{in}) \in K^n$ for $1 \leq i \leq t$. For each j with $1 \leq j \leq n$ we consider the polynomial

$$g_j(x_j) = (x_j - a_{1j}) \cdots (x_j - a_{tj}) \in K[x_j].$$

It is clear that $g_j(\mathbf{a}) = 0$ for all $\mathbf{a} \in \mathcal{V}(I)$, that is, $g_j \in \mathcal{I}(\mathcal{V}(I))$. By using the Strong Hilbert's Nullstellensatz, it follows that $g_j \in \sqrt{I}$ for all j. Therefore, for every $1 \leq j \leq n$, there exists an integer $m_j \geq 1$ such that $g_j^{m_j} \in I$. This implies, in turn, that $x_j^{tm_j} \in \mathtt{in}_<(I)$ for all j and for any monomial order $<$ on S. This proves (e).

To end the theorem's proof, we show (c) $\Rightarrow$ (b). If $\dim_K(S/I) = 0$, then $I = (1)$ and there is no maximal ideal of S which contains I. Let $\dim_K(S/I) = t \geq 1$ and assume that there are infinitely many maximal ideals which contain I. Then one may choose $t+1$ such distinct maximal ideals $\mathfrak{m}_1, \ldots, \mathfrak{m}_t$. We then have $I \subset \mathfrak{m}_1 \cap \cdots \cap \mathfrak{m}_{t+1}$. This implies that there exists a canonical surjective K-morphism

$$\varphi : S/I \longrightarrow S/\mathfrak{m}_1 \cap \cdots \cap \mathfrak{m}_{t+1} \cong \prod_{i=1}^{t+1}(S/\mathfrak{m}_i).$$

We obtain

$$t = \dim_K(S/I) \geq \dim_K(\prod_{i=1}^{t+1}(S/\mathfrak{m}_i)) = t+1,$$

contradiction. □

Let $I = (f_1, \ldots, f_r) \subset S$ be a zero-dimensional ideal. One may reduce the problem of solving the system of polynomial equations $f_1 = 0, \ldots, f_r = 0$ to solving n univariate equations. Unfortunately, there is no general algorithm to find roots of polynomials in one variable. However, in some concrete cases, by using, for instance, radical formulas, we are able to successfully apply this reduction.

For example, let us consider the system defined by the polynomials $f_1, f_2, f_3 \in \mathbb{C}[x, y, z]$, $f_1 = x^2+y^2+z^2-1$, $f_2 = x^2-y+z^2$, $f_3 = x-z$, and let $I = (f_1, f_2, f_3)$. Computing the Gröbner basis $\mathcal{G}$ of I with respect to the pure lexicographic order with $x > y > z$, we get $\mathcal{G} = \{x-z, y-2z^2, 4z^4+2z^2-1\}$. We first solve the equation in z and get four solutions. Next we substitute these solutions in the equation $y - 2z^2 = 0$, in order to get the second component, and, finally, from $x - z = 0$ we obtain the first component.

The previous example illustrates a general procedure to solve a polynomial system by using lexicographic Gröbner bases. Let $f_1 = 0, \ldots, f_r = 0$ be an algebraic system with finite and nonempty solution set. One computes the reduced Gröbner basis $\mathcal{G}$ with respect to the pure lexicographic order. Let $\mathcal{G} = \{g_1, \ldots, g_s\}$. By condition (e) in Theorem 3.22, we may assume that for any $1 \leq i \leq n$, $\mathtt{in}_{\text{plex}}(g_i) = x_i^{\nu_i}$ for some $\nu_i \geq 1$. In particular, we have $s \geq n$. Then $g_n \in K[x_n]$, $g_{n-1} \in K[x_{n-1}, x_n], \ldots, g_1 \in K[x_1, \ldots, x_n]$. Assuming that one may solve the univariate equation $g_n(x_n) = 0$, next we substitute each of these solutions in $g(x_{n-1}, x_n) = 0$ and get univariate equations in x_{n-1}, and so on. Finally, one has to keep only those solutions which vanish all the remaining polynomials $g_{n+1}, \ldots, g_s$.

Remark 3.23. The proof of the implication (c) $\Rightarrow$ (b) in Theorem 3.22 provides an upper bound for the cardinality of $\mathcal{V}(I)$ if I is a zero-dimensional ideal. Namely, we have

$$|\mathcal{V}(I)| \leq \dim_K(S/I) = \dim_K(S/\,\mathtt{in}_<(I))$$

for any monomial order $<$ on S, the equality being true by Macaulay's Theorem.

We give an example where the above upper bound for the cardinality of $\mathcal{V}(I)$ is sharp and another one which shows that this upper bound may be very large compared with the cardinality of $\mathcal{V}(I)$.

Let $I = (x^2 + y^2 + z^2 - 1, x^2 - y + z^2, x - z) \subset \mathbb{C}[x, y, z]$ be the ideal which we have considered after the proof of Theorem 3.22. With respect to the pure lexicographic order with $x > y > z$, I has the Gröbner basis $\mathcal{G} = \{x - z, y - 2z^2, 4z^4 + 2z^2 - 1\}$. By condition (e) in Theorem 3.22, it follows that $\mathcal{V}(I)$ is a finite set. Moreover, $\mathtt{in}_{\text{plex}}(I) = (x, y, z^4)$, whence $\dim_K(S/\,\mathtt{in}_{\text{plex}}(I)) = 4$. Note that $\mathcal{V}(I)$ has 4 elements, as well. Indeed, the polynomial $4z^4 + 2z^2 - 1$ has 4 distinct complex roots and each of them uniquely determines a point of $\mathcal{V}(I)$. Thus we have equality in the inequality from Remark 3.23.

However, the upper bound provided by $\dim_K(S/\,\mathtt{in}_<(I))$ may be very large compared with the cardinality of $\mathcal{V}(I)$. Let us take, for instance, the ideal $I = (x^3 - y^2z, y^2 - xz, xy + y^2 + z^2) \subset \mathbb{C}[x, y, z]$. The reduced Gröbner basis of I with respect to the pure lexicographic order is

$$\begin{aligned} \mathcal{G} \;&=\; \{x^3 - y^2z, xy + y^2 + z^2, xz - y^2, y^3 + y^2z + z^3, 3y^2z^2 + 6z^4, \\ &\quad\; yz^3 + 2z^4, z^5\}. \end{aligned}$$

Thus $\mathtt{in}_{\text{plex}}(I) = (x^3, xy, xz, y^3, y^2z^2, yz^3, z^5)$ and we have

$$\text{Mon}(\mathbb{C}[x, y, z]) \setminus \text{Mon}(\mathtt{in}_{\text{plex}}(I)) = \{1, x, y, z, x^2, y^2, yz, z^2, y^2z, yz^2, z^3, z^4\}.$$

Therefore $\dim_K(S/\,\mathtt{in}_{\text{plex}}(I)) = 12$, while $\mathcal{V}(I)$ is a singleton, namely $\mathcal{V}(I) = \{0\}$.

3.4. Ideals of initial forms

Given a monomial order $<$ on S, $\mathtt{in}_<(I)$ is the monomial ideal generated by the leading monomials of the polynomials of I with respect to the given order. In case that the monomial order is graded, then the leading monomial of any polynomial f is of maximal degree among all monomials of the support of f. However, for an ideal $I \subset S$, it is also interesting to study the so-called ideal of initial forms of I.

Let $f = f_0 + f_1 + \cdots + f_d \in S$ be a nonzero polynomial where f_i is the i-degree homogeneous component of f, and let $j = \min\{i : f_i \neq 0\}$. f_j is called the **initial form** of f and is denoted $\mathtt{In}(f)$. For example, if $f = xy + y^2 - y^3 + x^2y^2 \in K[x, y]$, then $\mathtt{In}(f) = xy + y^2$. We make the convention that $\mathtt{In}(0) = 0$.

Definition 3.24. *Let $I \subset S$ be an ideal. The* ***ideal of initial forms*** *of I is $\mathtt{In}(I) = (\mathtt{In}(f) : f \in I)$.*

The interest in studying ideals of initial forms comes from the fact that they are strongly related to tangent cones of varieties. If $\mathcal{V} = \mathcal{V}(I) \subset K^n$ is an affine algebraic variety which contains the origin, then the **tangent cone** of $\mathcal{V}$ at 0 is

$$C_0(\mathcal{V}) = \mathcal{V}(\mathtt{In}(f) : f \in \mathcal{I}(\mathcal{V})).$$

When the field K is algebraically closed, the tangent cone $C_0(\mathcal{V})$ is determined by the equations of $\mathtt{In}(I)$. Indeed, to prove this claim one has to show that $\mathcal{V}(\mathtt{In}(f) : f \in \mathcal{I}(\mathcal{V}(I))) = \mathcal{V}(\mathtt{In}(I))$, which, by Theorem 3.18, is equivalent to showing that $\mathcal{V}(\mathtt{In}(\sqrt{I})) = \mathcal{V}(\mathtt{In}(I))$. Since $I \subset \sqrt{I}$, it follows that $\mathtt{In}(I) \subset \mathtt{In}(\sqrt{I})$, thus $\mathcal{V}(\mathtt{In}(\sqrt{I})) \subset \mathcal{V}(\mathtt{In}(I))$. For the other inclusion it is enough to note that, by Problem 3.16, we have $\mathtt{In}(\sqrt{I}) \subset \sqrt{\mathtt{In}(I)}$. It follows that $\mathcal{V}(\mathtt{In}(\sqrt{I})) \supset \mathcal{V}(\sqrt{\mathtt{In}(I)}) = \mathcal{V}(\mathtt{In}(I))$.

In general, given an ideal $I \subset S$ generated by $f_1, \ldots, f_s$, it does not follow that $\mathtt{In}(I)$ is generated by the initial forms of $f_1, \ldots, f_s$. For example, let $I = (f_1, f_2) \subset K[x, y, z]$ where $f_1 = xy + y^2 - y^3$ and $f_2 = x^2 - y^2 + z^3$. Then $g = xy^3 - y^4 + yz^3 = (y - x)f_1 + yf_2 \in I$, hence $\mathtt{In}(g) \in \mathtt{In}(I)$. But $\mathtt{In}(g) = g$ and it is easily seen that $g \notin (\mathtt{In}(f_1), \mathtt{In}(f_2)) = (xy + y^2, x^2 - y^2)$.

How can we get a system of generators for $\mathtt{In}(I)$? At least two methods are known. We may either use local monomial orders which are basically designed to allow calculations in localizations of polynomial rings, or use Gröbner basis for homogenizations of ideals. We are going to explain here the later procedure. For the readers who are interested in studying local monomial orders we refer to the book [**GP02**].

Let $I \subset S$ be an ideal and $I^h \subset S[t]$ its homogenization with respect to the new indeterminate t over K. Let $<$ be a monomial order on $S[t]$ which has the property that among monomials of same degree, any monomial which is divisible by t is greater than any monomial in S. For example, the lexicographic order induced by $t > x_1 > \cdots > x_n$ satisfies this condition.

Proposition 3.25. *Let $I \subset S$ be an ideal, $I^h \subset S[t]$, and $\mathcal{G}$ a Gröbner basis of I^h with respect to $<$. Then $\mathtt{In}(I) = (\mathtt{In}(\bar{g}) : g \in \mathcal{G})$, where $\bar{g}$ is the dehomogenization of g.*

Proof. Let $f \in I$. We assume that for all polynomials $h \in I$ with the property that $\mathtt{in}_<(\mathtt{In}(h)) < \mathtt{in}_<(\mathtt{In}(f))$, we have $\mathtt{In}(h) \in (\mathtt{In}(\bar{g}), g \in \mathcal{G})$, and show that $\mathtt{In}(f) \in (\mathtt{In}(\bar{g}), g \in \mathcal{G})$. Let $f = f_i + \cdots + f_{d-1} + f_d$ where $f_i = \mathtt{In}(f)$. Then $f^h = t^{d-i} f_i + \cdots + t f_{d-1} + f_d$. We obviously have $\mathtt{in}_<(f^h) = t^{d-i}\,\mathtt{in}_<(\mathtt{In}(f))$. Since $\mathcal{G}$ is Gröbner basis of I^h, there exists $g \in \mathcal{G}$ such that $\mathtt{in}_<(g) \mid \mathtt{in}_<(f^h)$. On the other hand, we have $\mathtt{in}_<(g) = t^e\,\mathtt{in}_<(\mathtt{In}(\bar{g}))$ for some $e \geq 0$. Therefore, we get $t^e\,\mathtt{in}_<(\mathtt{In}(\bar{g})) \mid t^{d-i}\,\mathtt{in}_<(\mathtt{In}(f))$. As $\mathtt{in}_<(\mathtt{In}(\bar{g}))$ and $\mathtt{in}_<(\mathtt{In}(f))$ are monomials in S, it follows that $\mathtt{in}_<(\mathtt{In}(\bar{g})) \mid \mathtt{in}_<(\mathtt{In}(f))$. Then we may find $c \in K$, $c \neq 0$, and a monomial $\mathbf{x}^{\mathbf{a}} \in S$ such that the leading term of $\mathtt{In}(f)$ is equal to $c\mathbf{x}^{\mathbf{a}}\,\mathtt{in}_<(\mathtt{In}(\bar{g}))$. If $\mathtt{In}(f) = c\mathbf{x}^{\mathbf{a}}\,\mathtt{In}(\bar{g})$, then the proof is finished. Else, let $h = f - c\mathbf{x}^{\mathbf{a}}\bar{g}$. Then $\mathtt{In}(h) = \mathtt{In}(f) - c\mathbf{x}^{\mathbf{a}}\,\mathtt{In}(\bar{g})$ has the property that $\mathtt{in}_<(\mathtt{In}(h)) < \mathtt{in}_<(\mathtt{In}(f))$, thus, by induction, $\mathtt{In}(h) \in (\mathtt{In}(\bar{g}) : g \in \mathcal{G})$. Therefore, $\mathtt{In}(f) \in (\mathtt{In}(\bar{g}) : g \in \mathcal{G})$. □

Note that the set $\{\mathtt{In}(\bar{g}) : g \in \mathcal{G}\}$ is not necessarily a Gröbner basis of $\mathtt{In}(I)$ with respect to the induced order on S by $<$ (see Problem 3.18).

Corollary 3.26. *Let K be an algebraically closed field, $I \subset S$ an ideal, and $\mathcal{G}$ a Gröbner basis of I^h with respect to the order on $S[t]$ considered in the above proposition. Then $C_0(\mathcal{V}(I)) = \mathcal{V}(\mathtt{In}(\bar{g}) : g \in \mathcal{G})$.*

Problems

Problem 3.1. Find the generators for $I \cap K[x]$ and $I \cap K[y]$, where $I = (x^3 + x^2y - xy - y^2, xy - x - y + 1, x^2 - y^2) \subset K[x, y]$.

Problem 3.2. Let $<$ be a monomial order on S. Show that $<$ is an elimination order for $x_1, \ldots, x_t$ if and only if $x_i > x_j^m$ for all $1 \leq i \leq t$, $t+1 \leq j \leq n$, and $m \geq 0$.

Problem 3.3. Show that on the ring $K[x, y]$ there is exactly one elimination order for x.

Problem 3.4. Let $S = K[x_1, \ldots, x_n]$, $T = K[x_{t+1}, \ldots, x_n]$, $t \geq 1$, let $<$ be a monomial order on S, $I \subset S$ an ideal, and $\mathcal{G}$ a Gröbner basis of I with respect to $<$. We say that $\mathcal{G}$ has the property τ if for any $g \in \mathcal{G}$, $\mathtt{in}_<(g) \in T$ implies $g \in T$. Assume that $\mathcal{G}$ is a reduced Gröbner basis of I. Show that $\mathcal{G}$ has the property τ if and only if $\mathtt{in}_<(I) \cap T = \mathtt{in}_<(I \cap T)$.

Problem 3.5. Let T and S be as in the previous problem and $<$ a monomial order on S such that for any ideal $I \subset S$, one has $\mathtt{in}_<(I) \cap T = \mathtt{in}_<(I \cap T)$. Then $<$ is an elimination order on S for $x_1, \ldots, x_t$.

Problem 3.6. Let R be an arbitrary ring. An element $a \in R$ is called **nilpotent** if there exists an integer $m > 0$ such that $a^m = 0$.

Let $n \geq 1$ and $I = (x^n y^{n+1}, y^n z^{n+1}, z^n x^{n+1}) \subset K[x, y, z]$. Check whether the class of $f = xy + xz$ in $R = K[x, y, z]/I$ is a nilpotent element in R.

Problem 3.7. Let $f = x_1^3 + x_2^3 + x_3^3 - 3x_1x_2x_3$ and $I \subset \mathbb{Q}[x_1, x_2, x_3]$ be the ideal generated by f and its partial derivatives. Check whether $x_1x_2x_3$ is $\mathbb{Q}[x_1, x_2, x_3]/I$-regular.

Problem 3.8. Check whether $f_1 = x_1x_4 - x_2x_3$, $f_2 = x_1x_3 - x_2^2$ is a regular sequence on the factor ring $\mathbb{C}[x_1, x_2, x_3, x_4]/(x_2x_4 - x_3^2)$.

Problem 3.9. Compute a reduced Gröbner basis for the kernel of the K-algebra homomorphism $\varphi : K[x_1, x_2, x_3] \longrightarrow K[t]$ defined by $x_1 \mapsto t, x_2 \mapsto t^2, x_3 \mapsto t^3$.

Problem 3.10. Find the kernel of the morphism $\mathbb{Q}[x_1, x_2, x_3, x_4] \longrightarrow \mathbb{Q}[s, t]$ defined by $x_1 \mapsto s^4, x_2 \mapsto s^3t, x_3 \mapsto st^3, x_4 \mapsto t^4$.

Problem 3.11. Let $I \subset R = K[x_1, x_2]$, $I = (x_1, x_2)^2$, and $\mathcal{R}(I) = R[It] \subset K[x_1, x_2, t]$ the Rees ring of I. Let $\varphi : R[u_1, u_2, u_3] \longrightarrow \mathcal{R}(I)$ be the presentation of $\mathcal{R}(I)$ given by $u_1 \mapsto x_1^2t$, $u_2 \mapsto x_1x_2t$, and $u_3 \mapsto x_2^2t$. Compute a Gröbner basis for the kernel of φ.

Problem 3.12. Let $\mathbb{F} = \mathbb{Z}_3[x]/(x^3 - x - 1) = \mathbb{Z}_3(a)$ where a is the class of x modulo $(x^3 - x - 1)$. Find the minimal polynomial of $b = a^2 + a + 1 \in \mathbb{F}$ over $\mathbb{Z}_3$ by using a Gröbner basis of the ideal $(y - x^2 - x - 1, x^3 - x - 1) \subset \mathbb{Z}_3[x, y]$.

Problem 3.13. Let $f \in S$ be a polynomial of degree $d > 0$. Show that $f^h = t^d f(\frac{x_1}{t}, \ldots, \frac{x_n}{t})$.

Problem 3.14. Find a Gröbner basis of I^h for $I = (x_1x_2 - 1, x_1^2 - x_2) \subset K[x_1, x_2]$.

Problem 3.15. Let I, J be ideals in S. Show that:

(i) $(I + J)^h = I^h + J^h$.

(ii) $(I \cap J)^h = I^h \cap J^h$.

(iii) $(IJ)^h = I^h J^h$.

(iv) $(I\colon J)^h = I^h\colon J^h$.

(v) $(\sqrt{I})^h = \sqrt{I^h}$.

(vi) If $\mathfrak{p}$ is a prime ideal in S, then $\mathfrak{p}^h$ is a prime ideal in $S[t]$.

Problem 3.16. Let $I \subset S$ be an ideal. Show that $\mathtt{In}(\sqrt{I}) \subset \sqrt{\mathtt{In}(I)}$.

Problem 3.17. Let $I \subset S$ be an ideal, $I^h \subset S[t]$ its homogenization, and $g \in I^h$. Show that $\bar{g} \in I$ where $\bar{g}$ is the dehomogenization of g, that is, $\bar{g}$ is obtained from g by substituting t by 1.

Problem 3.18. Let $I = (xy + y^2 - y^3, x^2 - y^2 + z^3) \subset K[x, y, z]$.

(i) Compute a Gröbner basis $\mathcal{G}$ of $I^h \subset K[t, x, y, z]$ with respect to lexicographic order induced by $t > x > y > z$.

(ii) Show that the set $\{\bar{g} : g \in \mathcal{G}\}$ is not a Gröbner basis of $\mathtt{In}(I)$ with respect to lexicographic order.

Problem 3.19. Let $\eta_0 = x_1 + x_2 + x_3$, $\eta_1 = x_1\varepsilon + x_2\varepsilon^2 + x_3$, $\eta_2 = x_1\varepsilon^2 + x_2\varepsilon + x_3 \in \mathbb{C}[x_1, x_2, x_3]$, where $\varepsilon \in \mathbb{C}$ is a primitive 3-root of unity.

(i) Compute the reduced Gröbner basis of

$$J = (t_1 - \eta_0, t_2 - \eta_1, t_3 - \eta_2) \subset \mathbb{C}[x_1, x_2, x_3, t_1, t_2, t_3]$$

with respect to an elimination order for x_1, x_2, x_3.

(ii) Let $f \in \mathbb{C}[x_1, x_2, x_3]$ be a polynomial. Show that f is invariant to the cyclic permutation of variables if and only if

$$f = \sum_{\mathbf{a}=(a_0,a_1,a_2)} c_{\mathbf{a}} \eta_0^{a_0} \eta_1^{a_1} \eta_2^{a_2},$$

where $a_1 + 2a_2 \equiv 0 \bmod 3$ for every $\mathbf{a}$ such that $c_{\mathbf{a}} \neq 0$.

(iii) Express $f = x_1^2 x_2 + x_2^2 x_3 + x_3^2 x_1$ as a polynomial in η_0, η_1, η_2.

Problem 3.20. Let $I = (x^2 - y^3) \subset \mathbb{C}[x, y]$. Find an infinite family of maximal ideals which contain I.

Problem 3.21. (i) Use Gröbner bases to solve the system

$$\begin{cases} x^2 + 2y^2 - 2 & = 0, \\ x^2 + xy + y^2 - 2 & = 0, \end{cases}$$

over complex numbers.

(ii) Write $I = (x^2 + 2y^2 - 2, x^2 + xy + y^2 - 2) \subset \mathbb{R}[x, y]$ as an intersection of maximal ideals.

Problem 3.22. Find $\mathcal{V}(I) \subset \mathbb{C}^3$ for $I = (xz - y, xy + 2z^2, y - z) \subset \mathbb{C}[x, y, z]$.

Problem 3.23. Let $I \subset S$ be a zero-dimensional ideal. Prove that the following statements are equivalent:

(i) $|\mathcal{V}(I)| = \dim_K(S/I)$;

(ii) $I = \bigcap_{\mathbf{a} \in \mathcal{V}(I)} \mathfrak{m}_{\mathbf{a}}$;

(iii) I is a radical ideal.

Chapter 4

Gröbner bases for modules

It is not hard to extend the theory of Gröbner bases to submodules of a finitely generated free module over a polynomial ring. This more general concept allows us to compute the syzygy modules of finitely generated modules.

4.1. Modules

Let R be an arbitrary commutative ring. An **R-module** M is an abelian group together with a scalar multiplication $R\times M \to M$, $(a,m)\mapsto am$, such that the natural rules hold, namely: $1m = m$, $(ab)m = a(bm)$, $(a+b)m = am+bm$ and $a(m+n) = am+an$ for all $a,b\in R$ and all $m,n\in M$.

Observe that if R happens to be a field, then an R-module M is nothing else but an R-vector space. Other examples of modules are ideals. In particular, R itself may be viewed as an R-module.

Let M be an R-module. An abelian subgroup N of M is called a **submodule** of M, if $an\in N$ for all $a\in R$ and $n\in N$. A submodule of M is itself a module with addition and multiplication induced by that of M. The submodules of R are just its ideals.

Just as for ideals, one can perform several operations on submodules (see the problems at the end of this chapter).

Let $N\subset M$ be a submodule. We construct a new module, the so-called **factor module** M/N of M modulo N. Its elements are the sets $m+N = \{m+n\colon\ n\in N\}$. The element $m+N\in M/N$ is called the **residue class of m modulo N**, and m is called a **representative** of the residue class $m + N$. Observe that $m_1 + N = m_2 + N$ if and only if $m_1 - m_2 \in N$.

The module structure on M/N is defined by the following operations: $(m_1+N)+(m_2+N) = (m_1+m_2)+N$ and $a(m+M) = am+M$. One easily checks that the definition of the addition and scalar multiplication does not depend on the representatives of the residue classes.

As an example, consider an ideal $I \subset R$ and an R-module M. Then the set of all finite sums $\sum_i a_i m_i$ with $a_i \in I$ and $m_i \in M$ is a submodule of M, denoted IM. The factor module M/IM, which is an R-module, may as well be considered as an R/I-module with scalar multiplication defined by $(a + I)(m + IM) = am + IM$.

Let $\mathcal{G} \subset M$ be any subset. The submodule $N = (\mathcal{G})$ of M generated by $\mathcal{G}$ consists of all linear combinations of elements in $\mathcal{G}$, that is, of all finite sums of the form $\sum_{i=1}^{r} a_i m_i$ with $a_i \in R$ and $m_i \in \mathcal{G}$. If $\mathcal{G} = \emptyset$, then $N = 0$, by definition. We say that $\mathcal{G}$ is a **system of generators** of M, if $M = (\mathcal{G})$, and that M is **finitely generated**, if M admits a finite system of generators.

A system of generators $\mathcal{G}$ of M is called a **basis** of M, if each element of M can be *uniquely* expressed as a linear combination of elements in $\mathcal{G}$. In contrast to vector spaces, modules in general do not have a basis. For example, consider the ideal $I = (x, y)$ in the polynomial ring $S = K[x, y]$, and suppose that I has the basis $\mathcal{G}$. Then the cardinality of $\mathcal{G}$ is at least 2, because otherwise $x = gf$ and $y = hf$ for some $f \in I$ and $g, h \in K[x, y]$. This would imply that f divides $\gcd(x, y) = 1$, contradicting the fact that $f \in I$. Now let f_1, f_2 be two elements of $\mathcal{G}$. Then the element $f = f_1 f_2 \in I$, can be written in two ways as a linear combination of f_1 and f_2, namely as $f = f_1 f_2$ with $f_1 \in S$ and $f_2 \in \mathcal{G}$ or as $f = f_2 f_1$ with $f_2 \in S$ and $f_1 \in \mathcal{G}$.

A module F which admits a basis is called **free**. If F has a basis of cardinality n, then any other basis of F has cardinality n as well, as we shall see in the next lemma. We call the cardinality of a basis of F the **rank** of F, and denote it by $\operatorname{rank} F$.

Lemma 4.1. *Let F be a free module with a finite basis. Then all bases of F have the same cardinality.*

Proof. We choose a maximal ideal $\mathfrak{m} \subset R$. Then the residue classes of the elements of a basis of F form a basis of the $R/\mathfrak{m}$-module $F/\mathfrak{m}F$; see Problem 4.5. Since $R/\mathfrak{m}$ is a field, the $R/\mathfrak{m}$-module $F/\mathfrak{m}F$ is a vector space over this field. Hence all its bases have the same cardinality, and so all bases of F have the same cardinality. □

Let M and N be R-modules. A group homomorphism $\varphi \colon M \to N$ is called an **R-module homomorphism**, if $\varphi(am) = a\varphi(m)$ for all $a \in$

R and $m \in M$. The R-module homomorphism $\varphi\colon M \to N$ is called a **monomorphism**, if φ is injective, an **epimorphism**, if φ is surjective, and an **isomorphism**, if φ is bijective. We write $M \cong N$, if there exists an isomorphism $\varphi\colon M \to N$.

Let $\varphi\colon M \to N$ be an R-module homomorphism. The subset

$$\operatorname{Ker}(\varphi) = \{m \in M\colon\ \varphi(m) = 0\}$$

of M is called the **kernel** of φ. It is a submodule of M. Similarly, the **image** $\operatorname{Im}(\varphi)$ of φ is a submodule of N. Observe that $\varphi\colon M \to N$ is a monomorphism if and only if $\operatorname{Ker}(\varphi) = \{0\}$.

Theorem 4.2. *Let $\varphi\colon M \to N$ be an R-module homomorphism. Then $\operatorname{Im}(\varphi) \cong M/\operatorname{Ker}(\varphi)$.*

Proof. The map $\varphi'\colon M/\operatorname{Ker}(\varphi) \to \operatorname{Im}(\varphi)$, $m + \operatorname{Ker}(\varphi) \mapsto \varphi(m)$ is well defined and an R-module homomorphism. Obviously it is an epimorphism. Thus it remains to be shown that φ' is a monomorphism. Let $m + \operatorname{Ker}(\varphi) \in \operatorname{Ker}(\varphi')$. Then $\varphi(m) = 0$, so that $m \in \operatorname{Ker}(\varphi)$. This implies that $m + \operatorname{Ker}(\varphi) = 0 + \operatorname{Ker}(\varphi)$, as desired. $\square$

Corollary 4.3. *Let M be an R-module. Then there exists a free R-module F and a submodule $U \subset F$ such that $M \cong F/U$. The free R-module F can be chosen to be finitely generated, if M is finitely generated.*

Proof. Let $\mathcal{G}$ be a set of generators of M. Let F be the R-module of all sequences of elements of R indexed by $\mathcal{G}$, for which all but finitely many elements of the sequence are equal to zero. Thus an element of F is of the form $(a_g)_{g\in\mathcal{G}}$ with $a_g = 0$ for all but finitely many g. Addition and scalar multiplication is defined componentwise. For each $g \in \mathcal{G}$, let $e_g = (a_h)_{h\in\mathcal{G}}$ be the sequence with $a_h = 0$ for $h \neq g$, and $a_g = 1$. Then the elements $e_g \in F$ form a basis of F, and hence F is a free R-module. The map $\epsilon\colon\ F \to M$ with

$$\epsilon(\sum_{g\in\mathcal{G}} a_g e_g) = \sum_{g\in\mathcal{G}} a_g g$$

is an epimorphism of R-modules. Let $U = \operatorname{Ker}(\epsilon)$. Then Theorem 4.2 implies that $M \cong F/U$, as desired. $\square$

4.2. Monomial orders and initial modules

Let $S = K[x_1, \ldots, x_n]$ be the polynomial ring in n indeterminates over a field K. In the previous section we have seen that any finitely generated S-module M has a presentation F/U, where F is a finitely generated free module with a basis which has the same number of elements as the number of generators of M. The module U is called the **relation module** of M (with respect to this presentation).

We fix a basis $e_1, \dots, e_r$ of F. Our goal is to define the initial module of U. For this purpose we have to say what monomials in F are. We say that $m \in F$ is a **monomial**, if for some i, the element m is of the form ue_i, where u is a monomial in S. A submodule $U \subset F$ is called a **monomial module**, if it is generated by monomials. The following characterization of monomial modules is straightforward.

Proposition 4.4. *Let U be a submodule of the free S-module $F = \bigoplus_{j=1}^{r} Se_j$. Then U is a monomial module if and only if for each j there exist monomial ideals I_j such that $U = I_1e_1 \oplus I_2e_2 \oplus \cdots \oplus I_re_r$. In particular, U is finitely generated.*

Proof. Let U be a monomial module and let I_j be the monomial ideal which is generated by all monomials u for which ue_j is a generator of U. Then $U = I_1e_1 + I_2e_2 + \cdots + I_re_r$. That this sum is direct follows immediately from the fact that $e_1, \dots, e_r$ is a basis of F. □

A **monomial order** of the monomials of F is a total order $<$ satisfying the following two conditions:

(1) $m < um$ for all monomials $m \in F$ and all monomials $u \neq 1$ in S;

(2) if $m_1 < m_2$, then $um_1 < um_2$ for all monomials $m_1, m_2 \in F$ and all monomials $u \in S$.

Given a monomial order $<$ on S, there are two standard methods to define monomial orders on F. For $u, v \in \text{Mon}(S)$ and $i, j \in \{1, 2, \dots, r\}$, we define

Position over coefficient: $ue_i > ve_j$, if $i < j$ or $i = j$ and $u > v$;

Coefficient over position: $ue_i > ve_j$, if $u > v$ or $u = v$ and $i < j$.

For example, if $<$ is the lexicographic order on S and $F = Se_1 \oplus Se_2$, then $x_2e_1 > x_1e_2$, if the position is given more importance than the coefficient, and $x_1e_2 > x_2e_1$ in the opposite case.

We call the monomial order on F which is the (reverse) lexicographic order on the coefficients and gives priority to the position, the **(reverse) lexicographic order** on F.

In analogy to Proposition 2.2 we have

Proposition 4.5. *Let $<$ be a monomial order on F. Then any descending sequence $m_1 \geq m_2 \geq \cdots$ of monomials of F stabilizes.*

Proof. Let U be the submodule of F generated by the elements m_i. Since U is a monomial module, it follows from Proposition 4.4 that U is finitely generated. Thus there exist integers $i_1 < i_2 < \cdots < i_k$ such that the monomials $m_{i_1}, \ldots, m_{i_k}$ generate U. We claim that $m_i = m_{i_k}$ for all $i \geq i_k$. Indeed, let $i \geq i_k$. Then $m_i = um_{i_j}$ for some $u \in \mathrm{Mon}(S)$ and some j. Hence $m_{i_k} \geq m_i \geq m_{i_j} \geq m_{i_k}$, which implies that $m_i = m_{i_k}$. □

Given a monomial order on F, then for any element $f \in F$, the initial monomial, the leading coefficient and the leading term are defined in the same way as it is done for polynomials.

For example, if $f = (2x_2^2 - x_2x_3)e_1 + x_1^2e_2$, and $<$ denotes the lexicographic order on F, then $\mathtt{in}_<(f) = x_2^2e_1$, the leading coefficient is 2 and the leading term is $2x_2^2e_1$.

Let $U \subset F$ be a submodule of F. We let $\mathtt{in}_<(U)$ be the submodule of F which is generated by the monomials $\mathtt{in}_<(f)$ for *all* $f \in U$. The monomial module $\mathtt{in}_<(U)$ is called the **initial module** of U. Since $\mathtt{in}_<(U)$ is finitely generated, as we observed in Proposition 4.4, there exist elements $f_1, \ldots, f_m \in U$ such that $\mathtt{in}_<(U)$ is generated by $\mathtt{in}_<(f_1), \ldots, \mathtt{in}_<(f_m)$. Any such system of elements of U is called a **Gröbner basis** of U with respect to $<$.

Just as for ideals we have

Proposition 4.6. *Any Gröbner basis of U is a system of generators of U.*

By using Proposition 4.5, the proof of the preceding proposition is verbatim the same as that of Theorem 2.8.

Corollary 4.7. *Any submodule of a finitely generated free S-module is finitely generated.*

Remark 4.8. The statement in Corollary 4.7 is true for any Noetherian ring, as can be easily seen by induction on the rank of the free module. Indeed, let F be a free R-module with basis $e_1, \ldots, e_n$, and let $U \subset F$ be a submodule. Let $\pi\colon F \to G = Re_2 \oplus \cdots \oplus Re_n$ be the epimorphism given by $\pi(e_1) = 0$ and $\pi(e_i) = e_i$ for $i = 2, \ldots, n$, and let $\varphi\colon U \to G$ be the composition of the inclusion map $U \subset F$ with π. Then $\mathrm{Ker}(\varphi) = U \cap Re_1$ and $\mathrm{Im}(\varphi) = U/U \cap Re_1$. Since $Re_1 \cong R$, we may identify $U \cap Re_1$ with an ideal of R. Hence, since we assume that R is Noetherian, it follows that $U \cap Re_1$ is finitely generated. By the induction hypothesis, $U/U \cap Re_1$ is finitely generated as well, since it is a submodule of the free R-module G which has rank $n-1$. Since a system of generators of $U \cap Re_1$ together with the preimages in U of a system of generators of $U/U \cap Re_1$ generate U, the assertion follows.

4.3. The division algorithm and Buchberger's criterion and algorithm for modules

The division algorithm as well as Buchberger's criterion and algorithm as we know it for ideals have their complete analogue for modules. In this section we formulate the corresponding facts for modules, stress the differences but omit the proofs of the theorems because they are literally the same as in the case of ideals.

Let F be a free S-module with a fixed basis $e_1, \dots, e_r$. Then each element $f \in F$ has a unique presentation as a sum $f = \sum c_m m$ with $c_m \in S$ and $m \in \mathrm{Mon}(F)$, where $\mathrm{Mon}(F)$ denotes the set of monomials of F (with respect to the basis $e_1, \dots, e_r$). In this sum, all but finitely many of the coefficients are zero. We set $\mathrm{supp}(f) = \{m \in \mathrm{Mon}(F)\colon\ c_m \neq 0\}$.

Theorem 4.9. *Let f and $g_1, \dots, g_m$ be elements in F with $g_i \neq 0$. Given a monomial order $<$ on F, there exist polynomials $q_1, \dots, q_m$ and an element r in F with*

$$f = q_1 g_1 + q_2 g_2 + \cdots + q_m g_m + r$$

such that the following conditions are satisfied:

(i) *no element of* $\mathrm{supp}(r)$ *is contained in the monomial module* $(\mathtt{in}_<(g_1), \dots, \mathtt{in}_<(g_m))$;

(ii) $\mathtt{in}_<(f) \geq \mathtt{in}_<(q_i g_i)$ *for all i.*

An equation $f = q_1 g_1 + q_2 g_2 + \cdots + q_m g_m + r$ satisfying the conditions (i) and (ii) is called a **standard expression** of f, and r is called a **remainder** of f with respect to $g_1, \dots, g_m$. As in the case of polynomials we say that f **reduces to** 0 with respect to $g_1, \dots, g_m$, if f has a remainder zero with respect to $g_1, \dots, g_m$.

The proof of the theorem is very constructive and provides an algorithm to find a standard expression of f. We demonstrate this by an example.

Let $S = K[x_1, x_2, x_3]$, $F = Se_1 \oplus Se_2$, $f = (x_1^2 + x_1 x_2)e_1 - x_2 x_3 e_2$, $g_1 = x_1 e_1 - x_2 e_2$ and $g_2 = (x_3 - 1)e_2$. We choose the lexicographic order on F and want to compute a standard expression of f with respect to g_1 and g_2. To do this we proceed as in the case of polynomials:

$$\begin{aligned} f = h_0 &= x_1 g_1 + h_1, & h_1 &= x_1 x_2 e_1 + (x_1 x_2 - x_2 x_3)e_2,\\ h_1 &= x_2 g_1 + h_2, & h_2 &= (x_1 x_2 + x_2^2 - x_2 x_3)e_2,\\ h_2 &= -x_2 g_2 + h_3, & h_3 &= (x_1 x_2 + x_2^2 - x_2)e_2. \end{aligned}$$

Thus we obtain the standard expression

$$f = (x_1 + x_2)g_1 - x_2 g_2 + r \quad \text{with} \quad r = (x_1 x_2 + x_2^2 - x_2)e_2.$$

Just as for ideals we have

Proposition 4.10. *Let $<$ be a monomial order on the free S-module F, U a submodule of F, $g_1, \dots, g_m$ a Gröbner basis of U with respect to $<$, and $f \in F$. Then*

(a) *f has a unique remainder with respect to $g_1, \dots, g_m$.*

(b) *$f \in U$ if and only f reduces to 0 with respect to $g_1, \dots, g_m$.*

Next we want to generalize Buchberger's criterion and Buchberger's algorithm to modules. For this it is necessary to understand what the S-polynomials should be. Let F be a free S-module and $<$ a monomial order on F. The idea is the same as for ideals: for $f, g \in F$ we want to construct an element which is obtained as a linear combination of f and g such that their leading terms cancel. Say, $\mathtt{in}_<(f) = ue_i$ and $\mathtt{in}_<(g) = ve_j$. Obviously, if $i \neq j$, there is no linear combination of f and g such that the leading terms can cancel. Thus an analogue to S-polynomials can only be defined if $i = j$. In that case we set

$$S(f,g) = \frac{\operatorname{lcm}(u,v)}{cu} f - \frac{\operatorname{lcm}(u,v)}{dv} g, \tag{4.1}$$

where c is the coefficient of $\mathtt{in}_<(f)$ in f and d is the coefficient of $\mathtt{in}_<(g)$ in g. We call $S(f,g)$ the **S-element** of f and g.

With this definition the proof of the next theorem follows exactly the arguments in the proof of Theorem 2.14.

Theorem 4.11. *Let $U \subset F$ be a submodule of F and $\mathcal{G} = \{f_1, \dots, f_m\}$ a system of generators of U. Then $\mathcal{G}$ is a Gröbner basis of U if and only if for each pair (f_i, f_j) whose initial monomials involve the same basis element of F, the element $S(f_i, f_j)$ reduces to 0 with respect to $f_1, \dots, f_m$.*

As a result of this theorem one obtains an algorithm that allows us to compute the Gröbner basis of a submodule $U \subset F$. One starts with a system of generators $\mathcal{G} = \{f_1, \dots, f_m\}$ and computes the possible $S(f_i, f_j)$. If all $S(f_i, f_j)$ reduce to 0 with respect to $f_1, \dots, f_m$, then $\mathcal{G}$ is a Gröbner basis of U. Otherwise there is a nonzero remainder, which we call f_{m+1}. In that case we replace $\mathcal{G}$ by $\mathcal{G}' = \{f_1, \dots, f_{m+1}\}$ and proceed with $\mathcal{G}'$ as we did for $\mathcal{G}$. After a finite number of steps we arrive at a Gröbner basis of U.

We demonstrate the algorithm by an example. We let $S = K[x_1, x_2, x_3]$, $F = Se_1 \oplus Se_2$ and $U \subset F$ the submodule of F generated by

$$f_1 = (x_2 - x_3)e_1 + (x_1 + 1)e_2, \quad f_2 = (x_2 - 1)e_2, \quad f_3 = (x_2 - 1)e_1 + (x_3 - 1)e_2.$$

We want to compute a Gröbner basis of U with respect to the term order which gives priority to the coefficient before the position and orders the coefficients with respect to the pure lexicographic order. For this monomial order we have

$$\mathtt{in}_<(f_1) = x_1e_2, \quad \mathtt{in}_<(f_2) = x_2e_2, \quad \mathtt{in}_<(f_3) = x_2e_1.$$

The only S-element to be considered in the first step is that of f_1 and f_2. We compute its standard expression

$$\begin{aligned} S(f_1, f_2) &= x_2f_1 - x_1f_2 \\ &= f_1 + (-x_3 + 2)f_2 + (x_2 - x_3)f_3 + (x_3^2 - 2x_3 + 1)e_2. \end{aligned}$$

Thus a remainder of $S(f_1, f_2)$ with respect to f_1, f_2, f_3 is $f_4 = (x_3^2 - 2x_3 + 1)e_2$, which we add to the Gröbner basis of U.

In the next step we have to consider the standard expression of $S(f_1, f_4)$ and $S(f_2, f_4)$ with respect to f_1, f_2, f_3, f_4. We have

$$\begin{aligned} S(f_1, f_4) &= x_3^2f_1 - x_1f_4 \\ &= (2x_3 - 1)f_1 + (x_3^2 - 2x_3 + 1)f_3 + (-x_3 + 2)f_4 \\ &+ (-x_3^3 + 3x_3^2 - 3x_3 + 1)e_1 \end{aligned}$$

and

$$S(f_2, f_4) = (2x_3 - 1)f_2 - f_4.$$

Thus $S(f_1, f_4)$ has remainder $f_5 = (-x_3^3 + 3x_3^2 - 3x_3 + 1)e_1$, while $S(f_2, f_4)$ has remainder zero.

Finally, one checks that all remainders of the S-elements for f_1, f_2, f_3, f_4, and f_5 are zero with respect to f_1, f_2, f_3, f_4, f_5, so that by Theorem 4.11

$$f_1 = (x_2 - x_3)e_1 + (x_1 + 1)e_2, \; f_2 = (x_2 - 1)e_2, \; f_3 = (x_2 - 1)e_1 + (x_3 - 1)e_2,$$
$$f_4 = (x_3^2 - 2x_3 + 1)e_2, \; f_5 = (-x_3^3 + 3x_3^2 - 3x_3 + 1)e_1$$

is a Gröbner basis of U.

A reduced Gröbner basis of a submodule U of F is defined exactly as it is defined for ideals; see Definition 2.16. Just as in Theorem 2.17 one shows that a reduced Gröbner basis exists and is uniquely determined.

In the above example one obtains a reduced Gröbner basis by replacing f_1 by $f_1 - f_3$ and f_5 by $-f_5$.

4.4. Syzygies

4.4.1. How to compute syzygy modules. Let R be any Noetherian ring, and let M be a finitely generated R-module. By Corollary 4.3 and its proof, there exists a finitely generated free R-module F_0 and an epimorphism $\epsilon \colon F_0 \to M$. Let U_1 be the kernel of ϵ. According to Remark 4.8, U_1 is finitely generated, and hence as before, there exists a finitely generated free R-module F_1 and an epimorphism $\epsilon_1 \colon F_1 \to U_1$. Let $\varphi_1 \colon F_1 \to F_0$ be the composition of ϵ_1 with the inclusion map $U_1 \to F_0$. Then $\mathrm{Im}(\phi_1) = U_1 =$

$\operatorname{Ker}(\epsilon)$. Let $U_2 \subset F_1$ be the kernel of φ_1 (which is equal to the kernel of ϵ_1). Then, as before, there exists a finitely generated free R-module F_2 and an R-module homomorphism $\varphi_2\colon F_2 \to F_1$ such that $\operatorname{Im}(\varphi_2) = U_2 = \operatorname{Ker}(\varphi_1)$. Proceeding in this way we can construct a sequence of finitely generated free R-modules F_i and maps

$$(4.2) \qquad \cdots \xrightarrow{\varphi_3} F_2 \xrightarrow{\varphi_2} F_1 \xrightarrow{\varphi_1} F_0 \xrightarrow{\epsilon} M \longrightarrow 0,$$

such that ϵ is an epimorphism, $\operatorname{Ker}(\epsilon) = \operatorname{Im}(\varphi_1)$ and $\operatorname{Ker}(\varphi_i) = \operatorname{Im}(\varphi_{i+1})$ for all i.

More generally, a sequence of modules and module homomorphisms

$$\cdots \xrightarrow{\varphi_{i+1}} M_i \xrightarrow{\varphi_i} M_{i-1} \xrightarrow{\varphi_{i-1}} \cdots$$

is called **exact**, if $\operatorname{Ker}(\varphi_i) = \operatorname{Im}(\varphi_{i+1})$ for all i. Thus our sequence (4.2) is an exact sequence. Any such sequence where all F_i are free R-modules is called a **free R-resolution** of M, and for each i the module $U_i = \operatorname{Im}(\varphi_i)$ is called the **ith syzygy module** of M with respect to this resolution.

Observe that a free R-resolution of M is by no means unique. But we shall see in the next section that for graded modules there is a unique, up to isomorphism, "minimal" free R-resolution. For the moment we are only interested in constructing in a computational way a free resolution of a finitely generated module M over the polynomial ring $S = K[x_1, \ldots, x_n]$ which is given in the form $M = F/U$, where F is a finitely generated free S-module. We will see in this section that this can be done by using Gröbner bases.

Let F be a finitely generated free S-module, $U \subset F$ a submodule of F and $<$ a monomial order on F. Suppose further that $f_1, \ldots, f_m$ is a Gröbner basis of U. By Proposition 4.6 we know that $f_1, \ldots, f_m$ is a system of generators of U. We choose a free S-module G with basis $g_1, \ldots, g_m$, and let $\epsilon\colon G \to U$ be the epimorphism defined by $\epsilon(g_i) = f_i$ for $i = 1, \ldots, m$. The kernel of ϵ will be denoted by $\operatorname{Syz}(f_1, \ldots, f_m)$. Our task is to compute $\operatorname{Syz}(f_1, \ldots, f_m)$, which amounts to computing a system of generators of $\operatorname{Syz}(f_1, \ldots, f_m)$. The elements of $\operatorname{Syz}(f_1, \ldots, f_m)$ are called **relations** of U (with respect to the presentation $G \to U$). Notice that $\sum_{i=1}^m s_i g_i$ with $s_i \in S$ is a relation, if and only if $\sum_{i=1}^m s_i f_i = 0$.

Buchberger's criterion (see Theorem 4.11) gives us a set of relations for free. Indeed, for each pair f_i, f_j with $i < j$, whose initial monomials involve the same basis element of F, the element $S(f_i, f_j)$ reduces to zero with respect to $f_1, \ldots, f_m$. In other words, for each such pair we have an equation

$$(4.3) \qquad S(f_i, f_j) = q_{ij,1} f_1 + q_{ij,2} f_2 + \cdots + q_{ij,m} f_m,$$

which is a standard expression for $S(f_i, f_j)$. Recall from (4.1) that $S(f_i, f_j) = u_{ij}f_i - u_{ji}f_j$, where the terms u_{ij} and u_{ji} are chosen such that the leading terms of $u_{ij}f_i$ and $u_{ji}f_j$ are the same, so that they cancel in $S(f_i, f_j)$.

Equation (4.3) gives rise to the following relation:

$$(4.4) \qquad r_{ij} = u_{ij}g_i - u_{ji}g_j - q_{ij,1}g_1 - q_{ij,2}g_2 - \cdots - q_{ij,m}g_m.$$

Now we have

Theorem 4.12. *With the notation introduced, the relations r_{ij} arising from the S-elements of the Gröbner basis $f_1, \ldots, f_m$ of U generate* $\operatorname{Syz}(f_1, \ldots, f_m)$.

Proof. We let $V \subset \operatorname{Syz}(f_1, \ldots, f_m)$ be the submodule of $\operatorname{Syz}(f_1, \ldots, f_m)$ which is generated by the relations r_{ij}. We assign to each element $r = \sum_{j=1}^m h_j g_j$ in G the monomial $u_r = \max\{\mathtt{in}_<(h_j f_j)\colon j = 1, \ldots, m\}$. Assume now that r is a relation. We want to show that $r \in V$. We have that $u_r = w_r e_i$ for some i and $w_r \in \operatorname{Mon}(S)$. Without loss of generality, we may assume that $u_r = \mathtt{in}_<(h_j f_j)$ for $j = 1, \ldots, t$, that $\mathtt{in}_<(h_j f_j) < u_r$ for $j = t+1, \ldots, m$, and that the coefficient of $\mathtt{in}_<(f_j)$ is 1 for all j. Then for $j = 1, \ldots, t$, $\mathtt{in}_<(f_j) = u_j e_i$ for certain $u_j \in \operatorname{Mon}(S)$. For each j, let $\mathtt{in}_<(h_j) = a_j v_j$, where $a_j \in K$ and $v_j \in \operatorname{Mon}(S)$. Then

$$(4.5) \qquad w_r = v_j u_j \quad \text{for} \quad j = 1, \ldots, t, \quad \text{and} \quad \sum_{j=1}^t a_j v_j u_j = 0.$$

Therefore $v_1 u_1 = v_j u_j$ for $j = 2, \ldots, t$, so that there exist monomials w_j in S such that $v_j = w_j(\operatorname{lcm}(u_1, u_j)/u_j)$ for $j = 2, \ldots, t$.

Assume for the moment that we know already that $u_{r'} < u_r$ for the relation $r' = r + \sum_{j=2}^t a_j w_j r_{1j}$. By induction we may then assume that $r' \in V$, which then implies that $r \in V$, since $\sum_{j=2}^t a_j w_j r_{1j} \in V$. Thus it remains to be shown that indeed $u_{r'} < u_r$.

In fact, let $\sum_{k=1}^m q_{jk} f_k$ be the standard expression of $S(f_1, f_j)$ and set $\rho_j = \sum_{k=1}^m q_{jk} g_k$, then by (4.5) we get

$$\begin{aligned} r' &= \sum_{j=1}^t h_j g_j + \sum_{j=2}^t a_j w_j \Big(\frac{\operatorname{lcm}(u_1, u_j)}{u_1} g_1 - \frac{\operatorname{lcm}(u_1, u_j)}{u_j} g_j - \rho_j\Big) + \sum_{j=t+1}^m h_j g_j \\ &= \Big(h_1 + \sum_{j=2}^t a_j w_j \frac{\operatorname{lcm}(u_1, u_j)}{u_1}\Big) g_1 + \sum_{j=2}^t \Big(h_j - a_j w_j \frac{\operatorname{lcm}(u_1, u_j)}{u_j}\Big) g_j \\ &\quad - \sum_{j=2}^t a_j w_j \rho_j + \sum_{j=t+1}^m h_j g_j \end{aligned}$$

$$
\begin{aligned}
&= (h_1 - a_1v_1)g_1 + \sum_{j=2}^{t}(h_j - a_jv_j)g_j - \sum_{j=2}^{t} a_jw_j\rho_j + \sum_{j=t+1}^{m} h_jg_j \\
&= \sum_{j=1}^{t}(h_j - \mathtt{in}_<(h_j))g_j - \sum_{j=2}^{t} a_jw_j\rho_j + \sum_{j=t+1}^{m} h_jg_j.
\end{aligned}
$$

It follows that $u_{r'} < u_r$, since for each summand of the form hg_j in r' we have $\mathtt{in}_<(hg_j) < u_r$. This is obvious for the summands in $\sum_{j=1}^{t}(h_j - \mathtt{in}_<(h_j))g_j$ and the summands in $\sum_{j=t+1}^{m} h_jg_j$. The summands of $\sum_{j=2}^{t} a_jw_j\rho_j$ are $a_jw_jq_{jk}g_k$. Since $\sum_{k=1}^{m} q_{kj}f_k$ is the standard expression of $S(f_1, f_j)$, it follows that

$$
\begin{aligned}
\mathtt{in}_<(a_jw_jq_{jk}f_k) &\leq \mathtt{in}_<(a_jw_jS(f_1, f_j)) \\
&< \max\{\mathtt{in}_<(a_jw_j\tfrac{\operatorname{lcm}(u_1,u_j)}{u_1}f_1), \mathtt{in}_<(a_jw_j\tfrac{\operatorname{lcm}(u_1,u_j)}{u_j}f_j)\} \\
&\leq \max\{\mathtt{in}_<(\sum_{j=2}^{t} a_jw_j\tfrac{\operatorname{lcm}(u_1,u_j)}{u_1}f_1), \mathtt{in}_<(a_jw_j\tfrac{\operatorname{lcm}(u_1,u_j)}{u_j}f_j)\} \\
&= \max\{\mathtt{in}_<(h_1)\,\mathtt{in}_<(f_1), \mathtt{in}_<(h_j)\,\mathtt{in}_<(f_j)\} \\
&= \max\{\mathtt{in}_<(h_1f_1), \mathtt{in}_<(h_jf_j)\} = u_r.
\end{aligned}
$$

This completes the proof of the theorem. □

Corollary 4.13. *Let $U \subset F$ be a monomial submodule generated by the monomials $f_1, \ldots, f_m$. Then $\operatorname{Syz}(f_1, \ldots, f_m)$ is generated by the relations $r_{ij} = u_{ij}g_i - u_{ji}g_j$ for all $i < j$ for which f_i and f_j involve the same basis element. If $f_i = u_ie_k$ and $f_j = u_je_k$, then $u_{ij} = \operatorname{lcm}(u_i, u_j)/u_i$ and $u_{ji} = \operatorname{lcm}(u_i, u_j)/u_j$.*

Proof. Since the f_i are monomials, they form a Gröbner basis of U for any monomial order on F, and all S-elements reduce to zero. Thus the assertion follows from Theorem 4.12. □

Theorem 4.12 tells us how to compute a free S-resolution of the module $M = F/U$. In a first step we determine a Gröbner basis $f_1, \ldots, f_m$ of U by applying Buchberger's algorithm as described at the end of Section 4.3. In the course of applying this algorithm, the standard expressions for the S-elements $S(f_i, f_j)$ are computed. As a byproduct we obtain a generating set of relations for the elements $f_1, \ldots, f_m$ which are of the form r_{ij}, as described in (4.4). By Theorem 4.12, these relations generate the first syzygy module $\operatorname{Syz}(f_1, \ldots, f_m)$ of U. To obtain the next syzygy module, we apply the same procedure to $\operatorname{Syz}(f_1, \ldots, f_m)$. In this way, we obtain step by step a free S-resolution of M. As soon as a syzygy module turns out to be a free S-module, which can be tested by using Problem 4.11, we may end the construction of the free resolution at this step.

We demonstrate this procedure by the example given in Section 4.3: we let $S = K[x_1, x_2, x_3]$, $F = Se_1 \oplus Se_2$ and $U \subset F$ the submodule of F generated by

$$f_1 = (x_2 - x_3)e_1 + (x_1 + 1)e_2, \quad f_2 = (x_2 - 1)e_2, \quad f_3 = (x_2 - 1)e_1 + (x_3 - 1)e_2.$$

There we have shown that f_1, f_2, f_3 together with $f_4 = (x_3^2 - 2x_3 + 1)e_2$ and $f_5 = (-x_3^3 + 3x_3^2 - 3x_3 + 1)e_1$ is a Gröbner basis of U with respect to the monomial order which gives priority to the coefficient before the position and orders the coefficients with respect to the pure lexicographic order.

The S-elements of the f_i which involve the same basis elements have the following standard expressions:

$$\begin{aligned}
S(f_1, f_2) &= f_1 + (-x_3 + 2)f_2 + (x_2 - x_3)f_3 + f_4, \\
S(f_1, f_4) &= (2x_3 - 1)f_1 + (x_3^2 - 2x_3 + 1)f_3 + (-x_3 + 2)f_4 + f_5, \\
S(f_2, f_4) &= (2x_3 - 1)f_2 - f_4, \\
S(f_3, f_5) &= (3x_3^2 - 3x_3 + 1)f_3 + (x_3^2 - 2x_3 + 1)f_4 + f_5.
\end{aligned}$$

From the standard expression of $S(f_1, f_2)$ we obtain the equation

$$\begin{aligned}
0 &= x_2 f_1 - x_1 f_2 - (f_1 + (-x_3 + 2)f_2 + (x_2 - x_3)f_3 + f_4) \\
&= (x_2 - 1)f_1 + (-x_1 + x_3 - 2)f_2 + (-x_2 + x_3)f_3 - f_4,
\end{aligned}$$

which yields the relation

$$r_{12} = (x_2 - 1)g_1 + (-x_1 + x_3 - 2)g_2 + (-x_2 + x_3)g_3 - g_4.$$

Similarly, we obtain the relations

$$\begin{aligned}
r_{14} &= (x_3^3 - 2x_3 + 1)g_1 + (-x_3^2 + 2x_3 - 1)g_3 + (-x_1 + x_3 - 2)g_4 - g_5, \\
r_{24} &= (x_3^2 - 2x_3 + 1)g_2 + (-x_2 + 1)g_4, \\
r_{35} &= (-x_3^3 + 3x_3^2 - 3x_3 + 1)g_3 + (x_3^2 - 2x_3 + 1)g_4 + (-x_2 + 1)g_5.
\end{aligned}$$

According to Theorem 4.12 these relations generate $V = \operatorname{Syz}(f_1, f_2, \dots, f_5)$. Applying the assertion of Problem 4.11 to this example, we see that V is not a free module. Hence one would have to compute at least the next syzygy module to obtain a free resolution. But will one of the higher syzygy modules eventually be a free module, so that the resolution ends? This question will be discussed in Subsection 4.4.3

4.4.2. Systems of linear equations over the polynomial ring. Consider the system of linear equations

$$
\begin{aligned}
a_{11}y_1 + a_{12}y_2 + \cdots + a_{1s}y_s &= b_1 \\
a_{21}y_1 + a_{22}y_2 + \cdots + a_{2s}y_s &= b_2 \\
\vdots \qquad\qquad\quad \vdots \quad &\quad \vdots \\
a_{r1}y_1 + a_{r2}y_2 + \cdots + a_{rs}y_s &= b_r,
\end{aligned}
\tag{4.6}
$$

where the a_{ij} and b_k are elements of the polynomial ring $S = K[x_1, \ldots, x_n]$. An element $(h_1, \ldots, h_s) \in S^s$ is called a solution of (4.6) if $\sum_{j=1}^{s} a_{ij}h_j = b_i$ for $i = 1, \ldots, r$. We denote by $\mathcal{L}$ the set of all solutions of (4.6). If we replace in (4.6) all b_i by 0, then we obtain the **homogeneous** system of linear equations associated to (4.6). Its set of solutions will be denoted by $\mathcal{L}_0$.

We will now show how $\mathcal{L}$ can be computed. From a theoretical point of view this is easy. Let G be the free module S^s with canonical basis $g_1, \ldots, g_s$, and F the free module S^r with the canonical basis $f_1, \ldots, f_r$.

1. *Existence of a solution.* Set $a_j = \sum_{i=1}^{r} a_{ij}f_i$ for $j = 1, \ldots, s$ and $b = \sum_{i=1}^{r} b_i f_i$, and let $U \subset F$ be the submodule generated by the elements $a_1, \ldots, a_s$. Then $\mathcal{L} \neq \emptyset$ if and only if $b \in U$.

Indeed, if $(h_1, \ldots, h_s) \in \mathcal{L}$, then

$$
\begin{aligned}
b &= \sum_{i=1}^{r} b_i f_i = \sum_{i=1}^{r}(\sum_{j=1}^{s} a_{ij}h_j)f_i \\
&= \sum_{j=1}^{s} h_j(\sum_{i=1}^{r} a_{ij}f_i) = \sum_{j=1}^{s} h_j a_j.
\end{aligned}
$$

Reading these equalities backwards we see that if $b = \sum_{j=1}^{s} h_j a_j$, then $(h_1, \ldots, h_s) \in \mathcal{L}$.

2. *Description of $\mathcal{L}$.* Let $\epsilon\colon G \to U$ be the epimorphism with $\epsilon(g_j) = a_j$ for $j = 1, \ldots, s$. Then $V = \operatorname{Syz}(a_1, \ldots, a_s)$ is the kernel of ϵ. Thus $h_1g_1 + h_2g_2 + \cdots + h_sg_s \in V$, if and only if $\sum_{j=1}^{s} h_j a_j = 0$. This is equivalent to saying that $\sum_{j=1}^{s} a_{ij}h_j = 0$ for $i = 1, \ldots, r$. In other words, the syzygies of $a_1, \ldots, a_s$ correspond to the elements of $\mathcal{L}_0$.

Now, if $\mathcal{L} \neq \emptyset$ and $(h_1, \ldots, h_s) \in \mathcal{L}$ is a **particular solution**, then

$$
\mathcal{L} = (h_1, \ldots, h_s) + \mathcal{L}_0 = \{(h_1 + h_1', \ldots, h_s + h_s')\colon \sum_{j=1}^{s} h_j' g_j \in V\}.
$$

It follows from this discussion that in order to describe the elements of $\mathcal{L}$ explicitly we have to proceed as follows:

(i) Decide whether $b \in U$. If yes, then $\mathcal{L} \neq \emptyset$;

(ii) If $b \in U$, then express b as a linear combination of the generators $a_1, \dots, a_s$. The coefficients of this linear combination give us a particular solution;

(iii) Compute a system of generators of $V = \mathrm{Syz}(a_1, \dots, a_s)$. Then any element in $\mathcal{L}$ can be expressed as a sum of the particular solution and a linear combination of the generators of V.

For Step (iii) we have to find an algorithm to compute $\mathrm{Syz}(a_1, \dots, a_s)$ for the given set of generators $a_1, \dots, a_s$ of $U \subset F$ (which is not necessarily a Gröbner basis of U, as is assumed in Theorem 4.12). The following two lemmata tell us how we can do this.

Lemma 4.14. *Let W be the submodule of $F \oplus G$ generated by the elements $a_j + g_j$ for $j = 1, \dots, s$. Then*

$$\mathrm{Syz}(a_1, \dots, a_s) = W \cap G.$$

Proof. Let $w \in W$; then $w = \sum_{j=1}^{s} h_j(a_j + g_j)$ for suitable $h_j \in S$. It follows that $w \in W \cap G$ if and only if $\sum_{j=1}^{s} h_j a_j = 0$, which is the case if and only if $w \in \mathrm{Syz}(a_1, \dots, a_s)$. □

The intersection $W \cap G$ can be easily computed by using Gröbner bases.

Lemma 4.15. *Let H be a free S-module with basis $e_1, \dots, e_n$ and W a submodule of H. Let $1 \leq m \leq n$ be an integer and G be the free submodule of H with basis $e_m, \dots, e_n$, and let $<$ be the lexicographic order on H with $e_1 > e_2 > \cdots > e_n$. Furthermore, let $\mathcal{G} = w_1, \dots, w_r$ be a Gröbner basis of W with respect to $<$. We may assume that $\mathtt{in}_<(w_i) \in G$ if and only if $i \in \{1, \dots, s\}$. Then $w_1, \dots, w_s$ is a Gröbner basis of $W \cap G$.*

Proof. Let $w \in W \cap G$. Then $w = \sum_{i=k}^{n} c_i e_i$ with $c_i \in S$, $c_k \neq 0$ and $k \geq m$. Since $\mathcal{G}$ is Gröbner basis of W, we have $\mathtt{in}_<(w) = u\,\mathtt{in}_<(w_j)$ for some j and some monomial u. Since $\mathtt{in}_<(w) = \mathtt{in}(c_k)e_k$ it follows that $\mathtt{in}_<(w_j) = \mathtt{in}_<(d_k)e_k$ for some nonzero polynomial $d_k \in S$. The definition of the lexicographic order implies that $w_j = \sum_{i=k}^{n} d_i e_i$ with certain polynomials $d_i \in S$. In particular, $w_j \in G$ and $j \leq s$. This proves the assertion. □

Now we are ready to describe the algorithm to compute the set of solutions $\mathcal{L}$ of the system (4.6) of linear equations.

For Step (iii) we proceed as described in Lemma 4.14 and Lemma 4.15. For the Steps (i) and (ii) we apply again Lemma 4.14 and Lemma 4.15 to first compute $\mathrm{Syz}(a_1, \dots, a_s, b)$. In other words, we compute the Gröbner basis

$\mathcal{G}'$ of $W' \subset F \oplus G'$, where G' is the free S-module with basis $g_1, g_2, \dots, g_{s+1}$ and where W' is generated by the elements $a_j + g_j$ for $j = 1, \dots, s$ and the element $b + g_{s+1}$. Let $w_1, \dots, w_t$ be those elements of $\mathcal{G}'$ with $\mathtt{in}_<(w_i) \in G'$. Then these elements form a Gröbner basis of $\mathrm{Syz}(a_1, \dots, a_s, b)$. Hence if $w_i = \sum_{j=1}^{s+1} h_{ij} g_j$, then $h_{i,s+1} b = -\sum_{j=1}^{s} h_{ij} a_j$. It follows that $b \in (a_1, \dots, a_s)$ if and only if one of the $h_{i,s+1}$ is a nonzero constant polynomial. If this is the case and, say, $h_{i,s+1} = c$ with $c \in K \setminus \{0\}$, then we get the following presentation of b as the linear combination of the a_j, namely

$$b = -c^{-1} h_{i1} a_1 - c^{-1} h_{i2} a_2 - \cdots - c^{-1} h_{is} a_s.$$

The following example demonstrates this algorithm. We want to find the set of solutions $\mathcal{L}$ of the system of linear equations

$$\begin{aligned} x_1 y_1 + x_2 y_2 + x_3 y_3 &= -x_1^2 + x_2^2 + x_3^2 \\ (x_2 + x_3) y_1 + (x_1 + x_3) y_2 + (x_1 + x_2) y_3 &= 2 x_2 x_3 \end{aligned}$$

with coefficients in $S = K[x_1, x_2, x_3]$.

Let $a_1 = x_1 e_1 + (x_2 + x_3) e_2$, $a_2 = x_2 e_1 + (x_1 + x_3) e_2$, $a_3 = x_3 e_1 + (x_1 + x_2) e_2$ and $b = (-x_1^2 + x_2^2 + x_3^2) e_1 + 2 x_2 x_3 e_2$. For the Steps (i) and (ii) we have to compute (with respect to the lexicographic order) the Gröbner basis of the submodule $W' \subset \bigoplus_{i=1}^{6} S e_i$ generated by

$$a_1 + e_3, a_2 + e_4, a_3 + e_5, b + e_6.$$

The calculation shows that the Gröbner basis of W' consists of the above generators and the additional elements

$$\begin{aligned} (x_1^2 - x_2^2 - x_1 x_3 + x_2 x_3) e_4 + (x_1^2 + x_1 x_2 - x_2 x_3 + x_3^2) e_5 + (x_2 - x_3) e_6, \\ (x_2 - x_3) e_3 + (-x_1 + x_3) e_4 + (x_1 - x_2) e_5, \\ x_1 e_3 - x_2 e_4 - x_3 e_5 + e_6, \\ (x_1 x_2 + x_2^2 - x_1 x_3 - x_3^2) e_2 - x_3 e_4 + x_2 e_5, \\ (x_1^2 - x_2^2 + x_1 x_3 - x_2 x_3) e_2 - x_3 e_3 + x_3 e_4 + (x_1 - x_2) e_5. \end{aligned}$$

The element $x_1 e_3 - x_2 e_4 - x_3 e_5 + e_6$ tells us that the linear system of equations is solvable and that $(-x_1, x_2, x_3)$ is a particular solution.

For Step (iii) it is required to compute the Gröbner basis of the submodule $W \subset \bigoplus_{i=1}^{5} S e_i$ generated by

$$a_1 + e_3, a_2 + e_4, a_3 + e_5.$$

The Gröbner basis consists of these generators and the additional elements

$$\begin{aligned} (x_2 - x_3) e_3 + (-x_1 + x_3) e_4 + (x_1 - x_2) e_5, \\ (x_1 x_2 + x_2^2 - x_1 x_3 - x_3^2) e_2 - x_3 e_4 + x_2 e_5, \\ (x_1^2 - x_2^2 + x_1 x_3 - x_2 x_3) e_2 - x_3 e_3 + x_3 e_4 + (x_1 - x_2) e_5. \end{aligned}$$

From this we see that $\mathrm{Syz}(a_1, a_2, a_3)$ is generated by

$$(x_2 - x_3)e_3 + (-x_1 + x_3)e_4 + (x_1 - x_2)e_5.$$

Thus we obtain as the final result that the set of solutions of our linear system of equations is given by

$$\mathcal{L} = \{(-x_1, x_2, x_3) + f \cdot (x_2 - x_3, -x_1 + x_3, x_1 - x_2) \colon \; f \in S\}.$$

4.4.3. Schreyer's theorem. Our next goal is to show that each finitely generated S-module has a free resolution of length at most n, where n is the number of variables of the polynomial ring S. This is the celebrated **syzygy theorem** of Hilbert. We prove this theorem by using Gröbner bases following the arguments given by Schreyer [**Sc80**], who found this new proof of Hilbert's syzygy theorem. The essential idea is to choose suitable monomial orders in the computation of the syzygies.

Let F be a free S-module with basis $e_1, \ldots, e_r$ and $<$ a monomial order on F. Let $U \subset F$ be generated by $f_1, \ldots, f_m$, G a free S-module with basis $g_1, \ldots, g_m$, and $\epsilon \colon G \to U$ the epimorphism with $\epsilon(g_j) = f_j$ for $j = 1, \ldots, m$. We define a monomial order on G, again denoted $<$, as follows. Let ug_i and vg_j be monomials in G. Then we set

$$ug_i < vg_j \iff \mathtt{in}_<(uf_i) < \mathtt{in}_<(vf_j), \text{ or } \mathtt{in}_<(uf_i) = \mathtt{in}_<(vf_j) \text{ and } j < i.$$

Let us verify that $<$ is a monomial order on G. In order to see that $<$ is a total order on the monomials of G, we have to show that either $ug_i < vg_j$ or $ug_i \geq vg_j$.

Assume that we have $ug_i \not< vg_j$. Then $\mathtt{in}_<(uf_i) \not< \mathtt{in}_<(vf_j)$, and either $\mathtt{in}_<(uf_i) \neq \mathtt{in}_<(vf_j)$ or $j \geq i$. In the first case $\mathtt{in}_<(uf_i) > \mathtt{in}_<(vf_j)$, since $<$ is a total order on F. It follows in this case that $ug_i > vg_j$. In the second case $\mathtt{in}_<(uf_i) = \mathtt{in}_<(vf_j)$ and $j \geq i$. In this case $ug_i \geq vg_j$, by the definition of $<$ on G.

Next we check conditions (1) and (2) for monomial orders as defined before:

(1) Let $w \in \mathrm{Mon}(S)$, $w \neq 1$. Then $\mathtt{in}_<(uf_i) < w\,\mathtt{in}_<(uf_i) = \mathtt{in}_<(wuf_i)$, therefore $ug_i < wug_i$.

(2) Let $ug_i < vg_j$ and $w \in \mathrm{Mon}(S)$. If $\mathtt{in}_<(uf_i) < \mathtt{in}_<(vf_j)$, then $\mathtt{in}_<(wuf_i) = w\,\mathtt{in}_<(uf_i) < w\,\mathtt{in}_<(vf_j) = \mathtt{in}_<(wvf_j)$, and so $wug_i < wvg_j$. On the other hand, if $\mathtt{in}_<(uf_i) = \mathtt{in}_<(vf_j)$, then $j < i$ and $\mathtt{in}_<(wuf_i) = \mathtt{in}_<(wvf_j)$. So again, $wug_i < wvg_j$.

We call this monomial order defined on G the monomial order induced by $f_1, \ldots, f_m$ (and the monomial order $<$ on F).

The crucial result [**Sc80**] is now the following:

Theorem 4.16 (Schreyer). *Let F be a free S-module with basis $e_1, \dots, e_r$, and $<$ a monomial order on F. Let $U \subset F$ be a submodule of F with Gröbner basis $\mathcal{G} = \{f_1, \dots, f_m\}$. Then the relations r_{ij} arising from the S-elements of the f_i as described in (4.4) form a Gröbner basis of* $\operatorname{Syz}(f_1, \dots, f_m) \subset G$ *with respect to the monomial order induced by $f_1, \dots, f_m$. Moreover, one has*

$$\mathtt{in}_<(r_{ij}) = u_{ij} g_i,$$

where u_{ij} is defined as in (4.4).

Proof. Without loss of generality, we may assume that all the leading coefficients of the f_i are 1. Recall that $r_{ij} = u_{ij}g_i - u_{ji}g_j - q_{ij,1}g_1 - q_{ij,2}g_2 - \cdots - q_{ij,m}g_m$ where $q_{ij,1}f_1 + q_{ij,2}f_2 + \cdots + q_{ij,m}f_m$ is the standard expression for $S(f_i, f_j)$.

Since $\mathtt{in}_<(u_{ij}f_i) = \mathtt{in}_<(u_{ji}f_j)$ and since $i < j$ it follows that $\mathtt{in}_<(u_{ij}g_i - u_{ji}g_j) = u_{ij}g_i$. On the other hand, for all k we have the inequalities $\mathtt{in}_<(q_{ij,k}f_k) \leq \mathtt{in}_< S(f_i, f_j) < \mathtt{in}_<(u_{ij}f_i)$, so that $\mathtt{in}_<(q_{ij,k}g_k) < u_{ij}g_i$. Thus it follows that $\mathtt{in}_<(r_{ij}) = u_{ij}g_i$, as desired.

Next we show that the relations r_{ij} form a Gröbner basis of $V = \operatorname{Syz}(f_1, \dots, f_m)$. To this end, let $r = \sum_{j=1}^m r_j g_j \in V$ be an arbitrary relation. We have to show that there exists a relation r_{ij} such that $\mathtt{in}_<(r)$ is a multiple of $\mathtt{in}_<(r_{ij})$.

Let $\mathtt{in}_<(r_j g_j) = v_j g_j$ for $j = 1, \dots, m$, and let c_j be the coefficient of $\mathtt{in}(r_j g_j)$ in $r_j g_j$. Then $\mathtt{in}_<(r) = v_i g_i$ for some i. Now let $r' = \sum_j c_j v_j g_j$, where the sum is taken over the set $\mathcal{S}$ of those j for which $v_j\,\mathtt{in}_<(f_j) = v_i\,\mathtt{in}_<(f_i)$. Since we assume that $\mathtt{in}_<(r) = v_i g_i$, it follows that $j \geq i$ for all $j \in \mathcal{S}$. Substituting each g_j in r' by $\mathtt{in}_<(f_j)$, the sum becomes zero. Therefore r' is a relation of the elements $\mathtt{in}_<(f_j)$ with $j \in \mathcal{S}$. Hence by Corollary 4.13 the element r' is a linear combination of elements of the form $u_{kl}g_k - u_{lk}g_l$ with $k, l \in S$ and $k < l$. Since $j > i$ for all $j \in S$ with $j \neq i$, it follows at once that $\mathtt{in}_<(r')$ is a multiple of $u_{ij}g_i$ for some j. Since $\mathtt{in}_<(r) = \mathtt{in}_<(r')$ and $\mathtt{in}_<(r_{ij}) = u_{ij}g_i$, the assertion follows. ☐

The monomial order induced by $f_1, \dots, f_m$ allows some flexibility, since we are free to relabel the elements of the Gröbner basis as we want. Doing this in a clever way we obtain

Corollary 4.17. *With the notation introduced in Theorem 4.16, let the f_i be indexed in such way that whenever $\mathtt{in}_<(f_i)$ and $\mathtt{in}_<(f_j)$ for some $i < j$ involve the same basis element, say $\mathtt{in}_<(f_i) = ue_k$ and $\mathtt{in}_<(f_j) = ve_k$, then $u > v$ with respect to the pure lexicographic order induced by $x_1 > x_2 > \cdots > x_n$. Then it follows that if for some $t < n$ the variables $x_1, \dots, x_t$ do* *not appear in the initial monomial of the* f_j, *then the variables* $x_1, \ldots, x_{t+1}$ *do not appear in the initial monomial of the* r_{ij}.

Proof. By Theorem 4.16 we have $\mathtt{in}_<(r_{ij}) = (\mathrm{lcm}(u,v)/u)g_i$. Since $u > v$, and since u and v are monomials in the variables $x_{t+1}, \ldots, x_n$, it follows that the exponent of x_{t+1} in u is bigger than that of v. Thus $\mathrm{lcm}(u,v)/u$ is a monomial in the variables $x_{t+2}, \ldots, x_n$, as desired. □

As a consequence of Corollary 4.17 we finally obtain

Theorem 4.18 (Hilbert's syzygy theorem). *Let* M *be a finitely generated* S*-module over the polynomial ring* $S = K[x_1, \ldots, x_n]$. *Then* M *admits a free* S*-resolution*

$$0 \to F_p \longrightarrow F_{p-1} \longrightarrow \cdots \longrightarrow F_1 \longrightarrow F_0 \longrightarrow M \to 0$$

of length $p \leq n$.

Proof. Let $U \subset F$ be a submodule of the free S-module F with basis $e_1, \ldots, e_r$. Let $<$ be a monomial order on F, and $f_1, \ldots, f_m$ a Gröbner basis of U. Finally, let $t \leq n$ be the largest integer such that the variables $x_1, \ldots, x_t$ do not appear in any of the initial forms of the f_i. We prove by induction on $n-t$, that U has a free S-resolution of length $\leq \max\{0, n-t-1\}$

If $t \geq n-1$, then $\mathtt{in}_<(U) = \bigoplus_{j=1}^r I_j e_j$, where for each j, there exists a monomial ideal $J_j \subset K[x_n]$ such that $I_j = J_j S$. Since all monomial ideals in $K[x_n]$ are principal, it follows from Problem 4.11 that U is free.

If $t < n$, we may assume that the Gröbner basis $f_1, \ldots, f_m$ is labeled as described in Corollary 4.17. Then Theorem 4.16 together with Corollary 4.17 imply that $\mathrm{Syz}(f_1, \ldots, f_m)$ has a Gröbner basis with the property that the variables $x_1, \ldots, x_{t+1}$ do not appear in any of the leading monomials of the elements of the Gröbner basis. Thus, by induction, $\mathrm{Syz}(f_1, \ldots, f_m)$ has a free S-resolution of length $\leq n-t-2$. Composing this resolution with the exact sequence $0 \to \mathrm{Syz}(f_1, \ldots, f_m) \to G \to U \to 0$, we obtain for U a free S-resolution of length $\leq n-t-1$, as desired.

Now let M be an arbitrary finitely generated S-module. Then $M \cong F/U$, where F is a finitely generated free S-module. We may assume that $n > 0$. Then by the preceding arguments U has a free S-resolution of length $\leq n-1$. This implies that M has a free S-resolution of length $\leq n$. □

4.4.4. Graded rings and modules. In Chapter 1 we have seen that the polynomial ring $S = K[x_1, \ldots, x_n]$ has a decomposition $S = \bigoplus_{i \geq 0} S_i$ where for each i, S_i is the K-vector space of homogeneous polynomials of degree i. Observe that $S_i S_j \subseteq S_{i+j}$ for all i and j. Motivated by this example we introduce the following concept.

Definition 4.19. *Let K be a field. A ring R is called a **graded K-algebra**, if*

(i) $R = \bigoplus_{i \geq 0} R_i$, *where each R_i is a K-vector space;*

(ii) $R_0 = K$;

(iii) $R_i R_j \subset R_{i+j}$ *for all i, j.*

*The graded K-algebra is called **standard graded**, if $R = K[R_1]$ and* $\dim_K R_1 < \infty$.

Let R and R' be graded K-algebras. A K-algebra homomorphism $\varphi\colon R \to R'$ is called a **homomorphism of graded K-algebras** if $\varphi(R_i) \subset R'_i$ for all i.

Let $I \subset S$ be a graded ideal. Then $R = S/I$ is standard graded with $R_i = S_i/(I \cap S_i) = S_i/I_i$ for all i. Up to isomorphisms of graded K-algebras these are the only standard graded K-algebras as the following result shows.

Proposition 4.20. *Let R be a graded K-algebra with* $\dim_K R_1 = n$. *Then the following conditions are equivalent:*

(a) *R is standard graded.*

(b) *There exists a graded ideal $I \subset S = K[x_1, \ldots, x_n]$ and an isomorphism of graded K-algebras $R \cong S/I$.*

Proof. (a) $\Rightarrow$ (b): Let $r_1, \ldots, r_n$ be a K-basis of R_1, and define the K-algebra homomorphism $\varphi\colon S \to R$ by $\varphi(x_i) = r_i$ for $i = 1, \ldots, n$. Observe that $\varphi(S_i) = R_i$, since R is standard graded. Let $I \subset S$ be the kernel of φ. Since φ is surjective, φ induces a K-algebra isomorphism $S/I \cong R$.

Let $f \in I$, $f = \sum_i f_i$, where the f_i are the homogeneous components of f. Then $0 = \sum_i \varphi(f_i)$ with $\varphi(f_i) \in R_i$. Since $R = \bigoplus_{i \geq 0} R_i$ it follows that $\varphi(f_i) = 0$ for all i. In other words, $f_i \in I$ for all i, and hence by Proposition 1.2, I is a graded ideal. Since φ induces for all i an isomorphism $S_i/I_i \cong R_i$ of K-vector spaces, it follows that the isomorphism $S/I \cong R$ is an isomorphism of graded K-algebras.

The implication (b) $\Rightarrow$ (a) is obvious. $\square$

Let R be a graded K-algebra. An R-module M is called a **graded R-module** if $M = \bigoplus_{i \in \mathbb{Z}} M_i$ and $R_i M_j \subset M_{i+j}$. Since the homogeneous components of the elements of a system of generators of M is again a system of generators of M, we see that a finitely generated graded R-module can be generated by a finite system of homogeneous elements. Let α be the least degree of a generator in such a system, then $M_i = 0$ for all $i < \alpha$. We call α the **initial degree** of M.

Given a graded R-module M and $j \in \mathbb{Z}$, then $M(j)$ is defined to be the graded R-module whose graded components are $M(j)_i = M_{i+j}$ for all i.

A submodule U of the graded R-module M is called a **graded submodule** of M, if U is a graded R-module with $U_i = U \cap M_i$ for all i. If $U \subset M$ is a graded submodule of M, then M/U is a graded R-module with graded components M_i/U_i for all i.

An R-module homomorphism is called **homogeneous** if $\varphi(M_j) \subset N_j$ for all j. For example, for $f \in R_i$ the map $M(-i) \to M$ with $x \mapsto fx$ is a homogeneous R-module homomorphism. The kernel of a homogeneous R-module homomorphism $\varphi\colon M \to N$ is a graded submodule of M.

We denote by $\mathfrak{m} = (x_1, \ldots, x_n)$ the graded maximal ideal of the polynomial ring $S = K[x_1, \ldots, x_n]$. In the next section we shall need the following graded version of **Nakayama's lemma**.

Lemma 4.21. *Let M be a finitely generated graded S-module, and $m_1, \ldots, m_r$ homogeneous elements of M whose residue classes modulo $\mathfrak{m}M$ form a K-basis of $M/\mathfrak{m}M$. Then the elements $m_1, \ldots, m_r$ generate M.*

Proof. Let U be the graded submodule of M generated by $m_1, \ldots, m_r$. We want to show that $U = M$. Our hypothesis implies that $M = U + \mathfrak{m}M$. Let $m \in M$ be a homogeneous element. We will show that $m \in U$. To prove this we proceed by induction on the degree of m. We write $m = u + fn$ with homogeneous elements $u \in U$, $f \in \mathfrak{m}$ and $n \in M$, and such that $\deg m = \deg u = \deg fn$. If the degree of m coincides with the initial degree α of M, then $m = u$. If $\deg u > \alpha$, then either $n = 0$ and $m = u$, or $\deg n < \deg m$. In the second case we may assume by induction that $n \in U$, so that $m \in U$, as well. □

Corollary 4.22. *Let M be a finitely generated graded S-module. Then all homogeneous minimal systems of generators of M have the same cardinality, namely $\dim_K M/\mathfrak{m}M$.*

4.4.5. Graded free resolutions. Now let M be a finitely generated graded R-module, generated by the homogeneous elements $x_1, \ldots, x_r$ with $\deg x_i = a_i$ for all i, and let $F = \bigoplus_{i=1}^r Re_i$ be the graded free R-module with $\deg e_i = \deg x_i = a_i$ for all i. Then $F \cong \bigoplus_{i=1}^r R(-a_i)$, and the R-module homomorphism $\varphi\colon F \to M$ with $\varphi(e_i) = x_i$ for all i is homogeneous. Thus $U = \operatorname{Ker}(\varphi)$ is a graded submodule of F. In the case that R is Noetherian, the module U is finitely generated, and hence as before there exists a graded free R-module G and a surjective homogeneous R-module homomorphism $G \to U$. Composing this map with the inclusion map $U \subset F$ we obtain the exact sequence $G \to F \to M \to 0$ of graded R-modules. Proceeding in this way we obtain a graded free resolution, that is, a resolution of M

$$(4.7)\quad \cdots \xrightarrow{\varphi_3} F_2 \xrightarrow{\varphi_2} F_1 \xrightarrow{\varphi_1} F_0 \xrightarrow{\epsilon} M \longrightarrow 0,$$

where each F_i is a finitely generated graded free R-module and each φ_i as well as ϵ are homogeneous R-module homomorphisms.

In the following we restrict our attention to the case that $R = S = K[x_1, \dots, x_n]$. Let F be a finitely generated free graded S-module. We fix a homogeneous basis $e_1, \dots, e_n$ and a monomial order $<$ on F.

Proposition 4.23. *Let $U \subset F$ be a graded submodule of M. Then the reduced Gröbner basis of U consists of homogeneous elements.*

Proof. We apply Buchberger's algorithm to obtain a Gröbner basis of U. Starting with a homogeneous system of generators of U, one simply has to observe that the S-elements and remainders of homogeneous elements are again homogeneous. Thus the Gröbner basis $\mathcal{G}$ constructed in this way is homogeneous. The reduced Gröbner basis of U is essentially obtained from $\mathcal{G}$ by skipping superfluous elements and taking suitable remainders. Therefore the reduced Gröbner basis of U is also homogeneous. ☐

Corollary 4.24. *Let M be a finitely generated graded S-module. Then M admits a graded free resolution of length $\leq n$.*

Proof. It follows from Proposition 4.23 that the free resolution constructed in the proof of Theorem 4.18, which is based on Schreyer's theorem, yields a graded free resolution. ☐

The graded free resolution (4.7) is called **minimal**, if $\varphi(F_i) \subset \mathfrak{m}F_{i-1}$ for all i, where $\mathfrak{m} = (x_1, \dots, x_n)$ is the graded maximal ideal of S. This naming is justified by the fact, shown below, that the rank of the free modules in a graded minimal free resolution of a module is minimal compared with any other graded free resolution of the same module.

Theorem 4.25. *Let M be a finitely generated graded S-module. Then:*

(a) *M admits a minimal graded free resolution.*

(b) *Any two graded minimal free resolutions of M are isomorphic.*

Proof. (a) We choose a minimal set of homogeneous generators $m_1, \dots, m_r$ of M, and let F_0 be the graded free S-module $F_0 = \bigoplus_{i=1}^{r} Se_i$ with $\deg e_i = \deg m_i$ for $i = 1, \dots, r$. Let $\epsilon\colon F_0 \to M$ be the epimorphism of graded S-modules defined by $\epsilon(e_i) = m_i$ for all i. It follows from Lemma 4.21 that $\mathrm{Ker}(\epsilon) \subset \mathfrak{m}F_0$. The same argument shows that if we choose in each step of the construction of the resolution a minimal free presentation of the syzygies (that is, the homogeneous basis of the free modules maps onto a minimal system of generators of the syzygy module), then we obtain a minimal graded free resolution.

(b) We first claim: (∗) if $\varphi\colon N \to N'$ is an isomorphism of graded S-modules and $\epsilon\colon F \to N$ a minimal free presentation of N, $\epsilon'\colon F' \to N'$ a minimal free presentation of N', then there exists an isomorphism of graded S-modules $\psi\colon F \to F'$ such that $\epsilon' \circ \psi = \varphi \circ \epsilon$. In fact, let $e_1, \dots, e_r$ be a homogeneous basis of F. Choose homogeneous elements $f_1, \dots, f_r \in F'$ with $\epsilon'(f_i) = \varphi(\epsilon(e_i))$ for $i = 1, \dots, r$, and let $\psi\colon F \to F'$ be the homomorphism of graded S-modules with $\psi(e_i) = f_i$ for $i = 1, \dots, r$. Since $\operatorname{Ker} \epsilon \subset \mathfrak{m}F$ and $\operatorname{Ker} \epsilon' \subset \mathfrak{m}F'$, it follows that the induced homomorphisms $\bar{\epsilon}\colon F/\mathfrak{m}F \to N/\mathfrak{m}N$ and $\bar{\epsilon'}\colon F'/\mathfrak{m}F' \to N'/\mathfrak{m}N'$ are isomorphisms. This implies that the map $\bar{\psi}\colon F/\mathfrak{m}F \to F'/\mathfrak{m}F'$ is an isomorphism as well, because $\bar{\psi} = (\bar{\epsilon'})^{-1} \circ \bar{\varphi} \circ \bar{\epsilon}$. Finally, since F and F' are free S-modules, it follows that ψ is an isomorphism.

Now let

$$\cdots \xrightarrow{\varphi_3} F_2 \xrightarrow{\varphi_2} F_1 \xrightarrow{\varphi_1} F_0 \xrightarrow{\epsilon} M \longrightarrow 0$$

and

$$\cdots \xrightarrow{\varphi'_3} F'_2 \xrightarrow{\varphi'_2} F'_1 \xrightarrow{\varphi'_1} F'_0 \xrightarrow{\epsilon'} M \longrightarrow 0,$$

be two minimal graded free resolutions of M.

We want to show that there exist homogeneous isomorphisms $\psi_i\colon F_i \to F'_i$ such that $\epsilon' \circ \psi_0 = \epsilon$ and $\varphi'_i \circ \psi_i = \psi_{i-1} \circ \varphi_i$ for all $i > 0$. The existence of ψ_0 is guaranteed by (∗). Suppose we have already constructed $\psi_0, \dots, \psi_{i-1}$. Then it follows that $\operatorname{Im}(\varphi_i) \cong \operatorname{Im}(\varphi'_i)$, and we can again apply (∗). □

Let $\mathbb{F}$ be a graded minimal free resolution of M with $F_i = \bigoplus_j S(-j)^{\beta_{ij}(M)}$. The preceding theorem tells us that the numbers $\beta_{ij}(M)$ are uniquely determined by M. They are called the **graded Betti numbers** of M. We will see in the next subsection that they determine several important numerical invariants of M.

Theorem 4.26. *Let M be a finitely generated graded S-module, and $\mathbb{F}$ a graded free resolution of M with $F_i = \bigoplus_j S(-j)^{b_{ij}}$ for all i. Then*

$$\beta_{ij}(M) \le b_{ij}$$

for all i and j.

Proof. Assume $\mathbb{F}$ is a graded minimal free resolution of M. Then Theorem 4.25 implies that $\beta_{ij}(M) = b_{ij}$ for all i and j. Suppose now that $\mathbb{F}$ is not minimal. Then there exists $i > 0$ such that $\varphi_i(F_i) \not\subset \mathfrak{m}F_{i-1}$. Therefore there exists a homogeneous element $e \in F_i \setminus \mathfrak{m}F_i$ such that $f = \varphi_i(e) \in F_{i-1} \setminus \mathfrak{m}F_{i-1}$. Let $\mathbb{G}$ be the subcomplex of $\mathbb{F}$ with $G_j = 0$

if $j \neq i, i-1$, $G_i = Se$ and $G_{i-1} = Sf$. Then $\mathbb{G}$ is exact, and so it follows from the long exact homology sequence arising from the short exact sequence of complexes

$$0 \longrightarrow \mathbb{G} \longrightarrow \mathbb{F} \longrightarrow \mathbb{F}/\mathbb{G} \longrightarrow 0$$

that $\mathbb{F}' = \mathbb{F}/\mathbb{G}$ is acyclic with $H_0(\mathbb{F}') \cong M$; see [**E95**, A3.7]. Since $e \in F_i \backslash \mathfrak{m}F_i$ and $f \in F_{i-1}\backslash \mathfrak{m}F_{i-1}$, it follows that $F_i' = F_i/Se$ and $F_{i-1}' = F_{i-1}/Sf$ are graded free modules. Hence we see that $\mathbb{F}'$ is a graded minimal free resolution of M. Let $F_i' = \bigoplus_j S(-j)^{b_{ij}'}$ for all i and j. Since the rank of the free modules F_i' and F_{i-1}' is less than that of F_i and F_{i-1}, respectively, we may assume by induction that $\beta_{ij}(M) \leq b_{ij}'$ for all i and j, and since obviously $b_{ij}' \leq b_{ij}$ for all i and j, the desired conclusion follows. □

4.4.6. Numerical data arising from graded resolutions. In this section we explain how most of the important numerical invariants of a module can be deduced from the graded Betti numbers of M.

Let M be a finitely generated graded S-module with Betti numbers $\beta_{ij} = \beta_{ij}(M)$. The number

$$\operatorname{proj}\dim M = \max\{i\colon \ \beta_{ij} \neq 0 \quad \text{for some} \quad j\}$$

is called the **projective dimension** of M, and the number

$$\operatorname{reg} M = \max\{j\colon \ \beta_{i,i+j} \neq 0 \quad \text{for some} \quad i\}$$

is called the **regularity** of M.

Corollary 4.24 together with Theorem 4.26 implies that $\operatorname{proj}\dim M \leq n$. The **depth** of M is the length of a maximal homogeneous M-sequence. Recall that a sequence $\mathbf{x} = x_1, \dots, x_d$ of homogeneous elements of $\mathfrak{m}$ is called an **M-sequence** if the multiplication map $M/(x_1, \dots, x_{i-1})M \xrightarrow{x_i} M/(x_1, \dots, x_{i-1})M$ is injective for all i, and $M/(\mathbf{x})M \neq 0$.

By the theorem of Auslander-Buchsbaum (see [**BH98**, Section 1.3]) one has

$$\operatorname{depth} M + \operatorname{proj}\dim M = n,$$

so that

$$\operatorname{depth} M = \min\{i\colon \ \beta_{n-i,j} \neq 0 \quad \text{for some} \quad j\}.$$

Observe that in a minimal graded free resolution one always has

$$\max\{j\colon \beta_{ij} \neq 0\} \geq i$$

for all i. This implies that

$$\operatorname{reg} M \geq \alpha(M),$$

where $\alpha(M)$ is the least degree of a generator of M. We say that M has a **linear resolution**, if $\operatorname{reg} M = \alpha(M)$. This is the case if and only if all entries of matrices describing the differentials in the resolution are linear forms.

The numerical function $H(M,-)\colon \mathbb{Z} \to \mathbb{Z}_+$ with $H(M,i) = \dim_K M_i$ is called the **Hilbert function** of M, and the formal Laurent series

$$\mathrm{Hilb}_M(t) = \sum_i H(M,i)t^i$$

is called the **Hilbert series** of M. Note that $H(S,i) = \binom{n+i-1}{n-1}$, so that $\mathrm{Hilb}_S(t) = 1/(1-t)^n$.

Proposition 4.27. *Let M be a graded S-module with graded Betti numbers β_{ij}. Then*

$$\mathrm{Hilb}_M(t) = Q_M(t)/(1-t)^n \quad \text{with} \quad Q_M(t) = \sum_i (-1)^i (\sum_j \beta_{ij} t^j).$$

If we divide $Q_M(t)$ by the highest possible power of $1-t$, we obtain the presentation

$$\mathrm{Hilb}_M(t) = P_M(t)/(1-t)^d \quad \text{with} \quad P_M(1) \neq 0.$$

The number d in this presentation is the **Krull dimension**, and $P_M(1)$ is the **multiplicity** of M. The multiplicity, denoted $e(M)$, is a positive integer; see for example [**HH10**, Theorem 6.1.3]

Let $P_M(t) = \sum_{i=a}^b h_i t^i$. The coefficient vector $\mathbf{h} = (h_a, h_{a+1}, \ldots, h_b)$ is called the h-vector of M. Note that $a = \alpha(M)$. The a-**invariant** $a(M)$ of M is the degree of the rational function $\mathrm{Hilb}_M(t)$. In other words,

$$a(M) = b - d = \deg Q_M(t) - n \geq \max\{j\colon \ \beta_{ij} \neq 0 \quad \text{for some} \quad i\} - n.$$

It is convenient to display the graded Betti numbers of a module in a so-called **Betti diagram**; see Figure 1.

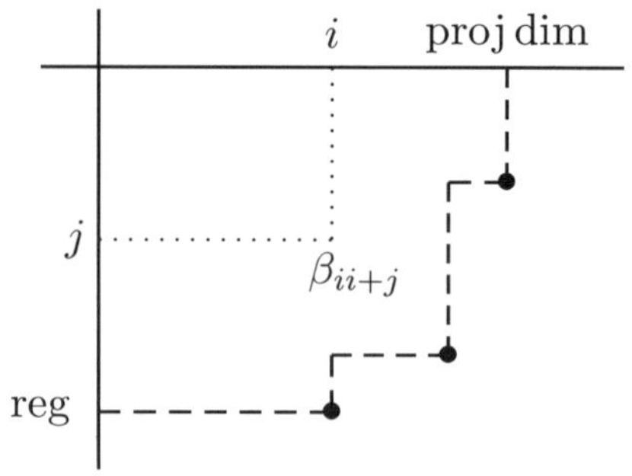

Figure 1

The graded Betti number $\beta_{i,i+j}$ is positioned at the coordinate (i, j). The nonzero graded Betti numbers are situated inside the area bounded by the dashed frame, where the corner points correspond to the nonzero graded Betti numbers, called the **extremal Betti numbers**. In the following example there are three extremal Betti numbers. They determine the regularity and projective dimension.

We conclude this section with a concrete example. Let

$$I = (x_1^2 - x_2x_3, x_3^2x_4, x_1x_2x_3, x_4^3) \subset S = K[x_1, x_2, x_3, x_4].$$

Then I has the minimal graded free resolution

$$0 \to S(-8) \to S^2(-6) \oplus S^3(-7) \to S^6(-5) \oplus S(-6)$$
$$\to S(-2) \oplus S^3(-3) \to I \to 0.$$

Thus the ideal I has the Betti diagram displayed in Figure 2.

I	0	1	2	3
2	1	–	–	–
3	3	–	–	–
4	–	6	2	–
5	–	1	3	1

Figure 2

From the Betti diagram we read off that $\operatorname{proj\,dim} I = 3$ and $\operatorname{reg} I = 5$. It follows that $\operatorname{proj\,dim} S/I = 4$, so that $\operatorname{depth} S/I = 0$.

For the Hilbert series of S/I we find

$$\operatorname{Hilb}_{S/I}(t) = \frac{1 - t^2 - 3t^3 + 6t^5 - t^6 - 3t^7 + t^8}{(1-t)^4} = \frac{1 + 3t + 5t^2 + 4t^3 - t^5}{1-t}.$$

From this we deduce that $\dim S/I = 1$, $e(S/I) = 12$ and $a(S/I) = 4$.

For any finitely generated graded S-module one has $\operatorname{depth} M \leq \dim M$.

Definition 4.28. *A finitely generated graded S-module M is called* ***Cohen-Macaulay*** *if* $\operatorname{depth} M = \dim M$.

By the Buchsbaum–Auslander theorem, $\operatorname{depth} M = \dim M$ if and only if $\operatorname{proj\,dim} M = \operatorname{codim} M$, where $\operatorname{codim} M$ is defined to be $\dim S - \dim M$. For a Cohen–Macaulay module the rank of the last free module in the minimal graded free resolution of M is called the **type** of M. A standard graded K-algebra $R = S/I$ is called Cohen–Macaulay, if R, viewed as an S-module, is Cohen–Macaulay, and it is called **Gorenstein**, if it is Cohen–Macaulay and the type of R is 1. It is known that R is Gorenstein if and only if R (as a module over itself) has finite injective dimension. For the proof of these basic facts we refer to [**BH98**]. Cohen–Macaulay rings play in important role in the homological theory of commutative rings. In the hierarchy of rings they are ranking right after regular rings and Gorenstein rings with regard to pleasant homological properties. Thus it is not a surprise that the Cohen–Macaulay property of rings or modules often reflect distinguished properties of the combinatorial objects to which they are attached, as exemplified in Stanley's famous proof of the upper bound theorem. Other instances can be found in the books [**S96**], [**HH10**], [**MS05**] and [**V90**].

We conclude this section by showing that Hilbert functions can be used to identify the initial ideal of a given ideal.

Proposition 4.29. *Let K be a field, $S = K[x_1, \ldots, x_n]$ the polynomial ring over K in the variables $x_1, \ldots, x_n$ and $I \subset S$ a graded ideal. We fix a monomial order $<$ and let $J \subset \mathtt{in}_<(I)$ be a monomial ideal. Then the following conditions are equivalent:*

(a) $J = \mathtt{in}_<(I)$;

(b) $\mathrm{Hilb}_{S/I}(t) = \mathrm{Hilb}_{S/J}(t)$.

Proof. Since $J \subset \mathtt{in}_<(I)$, it follows that $\mathrm{Hilb}_{S/\,\mathtt{in}_<(I)}(t) \leq \mathrm{Hilb}_{S/J}(t)$ coefficientwise, and equality holds if and only if $J = \mathtt{in}_<(I)$. Thus it remains to be shown that $\mathrm{Hilb}_{S/\,\mathtt{in}_<(I)}(t) = \mathrm{Hilb}_{S/I}(t)$. But this follows from Macaulay's theorem (Theorem 2.6). □

Corollary 4.30. *Let K be a field, $S = K[x_1, \ldots, x_n]$ the polynomial ring over K in the variables $x_1, \ldots, x_n$, $I \subset S$ a graded ideal and a monomial order $<$ on S. Then $\dim S/I = \dim S/\,\mathtt{in}_<(I)$.*

Proof. By Proposition 4.29 the graded rings S/I and $S/\,\mathtt{in}_<(I)$ have the same Hilbert function. Since the Hilbert function determines the Krull dimension, the assertion follows. □

4.4.7. $\mathbb{Z}^n$-graded modules. Monomial ideals in the polynomial ring $S = K[x_1, \ldots, x_n]$ are graded ideals. But in fact they have a finer graded structure as the one considered so far; their graded components can be indexed by $\mathbb{Z}^n$.

Definition 4.31. *Let K be a field. A ring R is called a **$\mathbb{Z}^n$-graded K-algebra**, if*

(i) $R = \bigoplus_{\mathbf{a} \in \mathbb{Z}^n} R_a$;

(ii) $R_0 = K$;

(iii) $R_{\mathbf{a}} R_{\mathbf{b}} \subset R_{\mathbf{a}+\mathbf{b}}$ *for all* $\mathbf{a}, \mathbf{b} \in \mathbb{Z}^n$.

Let R be a $\mathbb{Z}^n$-graded K-algebra. An R-module M is called $\mathbb{Z}^n$-graded, if $M = \bigoplus_{\mathbf{a} \in \mathbb{Z}^n} M_{\mathbf{a}}$, and $R_{\mathbf{a}} M_{\mathbf{b}} \subset M_{\mathbf{a}+\mathbf{b}}$ for all $\mathbf{a}, \mathbf{b} \in \mathbb{Z}^n$.

The polynomial ring is naturally $\mathbb{Z}^n$-graded with graded components

$$S_{\mathbf{a}} = K\mathbf{x}^{\mathbf{a}} \quad \text{if} \quad \mathbf{a} \in \mathbb{N}^n, \quad \text{and} \quad S_{\mathbf{a}} = 0 \quad \text{otherwise},$$

and each monomial ideal $I \subset S$ is in an obvious way a $\mathbb{Z}^n$-graded S-module. All basic properties of graded modules are valid similarly for a $\mathbb{Z}^n$-graded module. In particular, each finitely generated $\mathbb{Z}^n$-graded S-module admits a minimal $\mathbb{Z}^n$-graded free resolution.

Let F be a $\mathbb{Z}^n$-graded free S-module with homogeneous basis $e_1, \dots, e_m$ and $\deg e_i = \mathbf{a}_i$ for $i = 0, \dots, m$. Then $F_{\mathbf{a}}$ is the K-vector space spanned by all monomials $\mathbf{x}^{\mathbf{a}-\mathbf{a}_i} e_i$ for which $\mathbf{a} - \mathbf{a}_i \in \mathbb{N}^n$.

In the following we present results from [**FH11**] on initial modules of syzygies of $\mathbb{Z}^n$-graded modules which are similar in nature to Corollary 4.17 of the theorem of Schreyer. We fix a monomial order on S and let $<$ be the monomial order on F induced by the monomial order on S which gives priority to the position over the coefficients.

Let $M \subset F$ be a $\mathbb{Z}^n$-graded submodule. Then $\mathtt{in}_<(M)$ is generated by all elements $\mathtt{in}_<(u)$ where $u \in M$ is homogeneous. Let u be homogeneous of degree $\mathbf{a}$, say, $u = \sum_i c_i u_i e_i$ with $c_i \in K$, $u_i \in \mathrm{Mon}(S)$ and $\deg u_i + \deg e_i = \mathbf{a}$ for all i with $c_i \neq 0$. Then $\mathtt{in}_<(u) = u_j e_j$, where $j = \min\{i\colon\ c_i \neq 0\}$. Thus we see that $\mathtt{in}_<(M)$ depends only on the basis $\mathcal{F} = e_1, \dots, e_m$ of F and *not* on the given monomial order on S. Hence we denote the initial module of M by $\mathtt{in}_{\mathcal{F}}(M)$.

Our considerations so far can be summed up as follows:

Lemma 4.32. *With the assumptions and notation introduced we have*

$$\mathtt{in}_{\mathcal{F}}(M) = \bigoplus_{i=1}^{m} I_j e_j,$$

where $I_j \cong (M \cap \bigoplus_{k=j}^m Se_k)/(M \cap \bigoplus_{k=j+1}^m Se_k)$ *for* $j = 1, \dots, m$.

We call the basis $\mathcal{F} = e_1, \dots, e_m$ of F **lex-refined**, if $\deg(e_1) \geq \deg(e_2) \geq \dots \geq \deg(e_m)$ in the pure lexicographical order.

In the following we present a result which is a sort of analogue to the theorem of Schreyer. Let M be a $\mathbb{Z}^n$-graded S-module, and

$$\mathbb{F}\colon \quad \cdots \xrightarrow{\varphi_3} F_2 \xrightarrow{\varphi_2} F_1 \xrightarrow{\varphi_1} F_0 \xrightarrow{\epsilon} M \longrightarrow 0,$$

a $\mathbb{Z}^n$-graded free resolution of M. We set $Z_p(\mathbb{F}) = \mathrm{Im}(\varphi_p)$ for all p. Then $Z_p = Z_p(\mathbb{F})$ is the pth syzygy module of M with respect to the resolution $\mathbb{F}$.

Theorem 4.33. *Let* $1 \leq p \leq n$ *be an integer, and* $\mathcal{F}$ *a lex-refined basis of* F_{p-1}. *Then* $\mathtt{in}_{\mathcal{F}}(Z_p) = \bigoplus_{j=1}^m I_j e_j$, *where the minimal set of monomial generators of each* I_j *is contained in* $K[x_p, \dots, x_n]$.

Proof. The statement is trivial for $p = 1$. We may therefore assume that $p \geq 2$. Let $n \in Z_p$ be a homogeneous element of Z_p with $\mathtt{in}(n) = u_i e_i$ and such that u_i is a minimal generator of I_i. Let k be the smallest number such that x_k divides $\mathtt{in}(n) = u_i e_i$, and suppose that $k < p$.

Consider the exact sequence

$$0 \longrightarrow Z_p \longrightarrow F_{p-1} \xrightarrow{\varphi_{p-1}} F_{p-2} \longrightarrow Z_{p-2} \longrightarrow 0$$

where we set $Z_{p-2} = M$ if $p = 2$. According to Problem 4.6, $x_1, \ldots, x_{p-2}$ is a regular sequence on Z_{p-2}. We denote by "overline" the reduction modulo $(x_1, \ldots, x_{k-1})$. Then the above exact sequence yields the exact sequence

$$0 \longrightarrow \bar{Z}_p \longrightarrow \bar{F}_{p-1} \xrightarrow{\bar{\varphi}_{p-1}} \bar{F}_{p-2}.$$

Hence $\bar{Z}_p$ may be identified with its image in $\bar{F}_{p-1}$.

Thus $\bar{n}$ can be written as $\bar{n} = c_i u_i \bar{e}_i + c_{i+1} u_{i+1} \bar{e}_{i+1} + \cdots$ with $c_j \in K$, $u_j \in \operatorname{Mon}(S)$, and $c_i \neq 0$.

Since $u_j \in K[x_k, \ldots, x_n]$ for all j with $c_j \neq 0$ and since $\bar{n}$ is homogeneous, it follows that $\deg_t \bar{e}_j = \deg_t \bar{e}_i$ for all $t \leq k-1$ and all j with $c_j \neq 0$. (Here, for any homogeneous element r, we denote by $\deg_t r$ the tth component of $\deg r$.) Therefore, since x_k divides u_i, it follows that x_k divides $u_j \neq 0$ for $j > i$ with $c_j \neq 0$, because $\deg_k \bar{e}_i = \deg_k e_i \geq \deg_k e_j = \deg_k \bar{e}_j$ for $j > i$. This implies that x_k divides $\bar{n}$. Thus there exist $w \in \bar{F}_{p-1}$ such that $\bar{n} = x_k w$. It follows that $x_k \bar{\varphi}_{p-1}(w) = \bar{\varphi}_{p-1}(\bar{n}) = 0$. Since x_k is a nonzero divisor on $\bar{F}_{p-2}$, we see that $\bar{\varphi}_{p-1}(w) = 0$. This implies that $w \in \bar{Z}_p$. Let $m = d_r v_r e_r + \cdots + d_i v_i e_i + \cdots$ be a homogeneous element in F_{p-1} such that $\bar{m} = w$ with $v_j \in \operatorname{Mon}(S)$ and $d_j \in K$ for all j, and $d_r \neq 0$. Then $r \leq i$ and $u_i = x_k v_i$.

Suppose that $r < i$. Since $x_j \nmid u_i$ for all $j < k$, and since m is homogeneous it follows that

$$\deg_t v_r e_r = \deg_t v_i e_i = \deg_t e_i \quad \text{for all} \quad t < k. \tag{4.8}$$

On the other hand, since $\bar{n} = x_k \bar{m} = d_r x_k \bar{v}_r \bar{e}_r + \cdots$, we see that $x_k \bar{v}_r = 0$, and this implies that v_r is divisible by some x_j with $j < k$. Let s be the smallest such integer. Then by (4.8) we deduce that $\deg_j e_r = \deg_j e_i$ for $j < s$ and $\deg_s e_r < \deg_s e_i$. Hence $\deg e_r < \deg e_i$ (with respect to the pure lexicographic order), contradicting the choice of our basis. Thus $r = i$, and consequently, $v_i \in I_i$. But this is again a contradiction, since $u_i = x_k v_i$ and since u_i is a minimal generator of I_i. □

A vector $\mathbf{a} \in \mathbb{Z}^n$ is called **squarefree** if $a_i \in \{0, 1\}$ for all i. In the following we consider $\mathbb{Z}^n$-graded S-modules whose generators have squarefree degrees. Typical examples are squarefree monomial ideals.

Proposition 4.34. *Let M be a finitely generated $\mathbb{Z}^n$-graded S-module which admits a set of generators $m_1, \ldots, m_r$ whose degrees are squarefree, and let $\epsilon\colon \bigoplus_{i=1}^r Se_i \to M$ be the epimorphism of $\mathbb{Z}^n$-graded modules with $e_i \mapsto m_i$ for $i = 1, \ldots, r$. Then* $\operatorname{Ker}\epsilon$ *is also minimally generated by elements whose degrees are squarefree.*

Proof. Let $n = \sum_i c_i u_i e_i$ be a minimal homogeneous generator of $\operatorname{Ker}\epsilon$ of degree $\mathbf{a}$. Assume that $a_j > 1$ for some j. Then it follows that x_j divides u_i for all i with $c_i \neq 0$. This implies that x_j divides n, contradicting the fact that n is a minimal generator of $\operatorname{Ker}\epsilon$. □

Corollary 4.35. *Let M be a $\mathbb{Z}^n$-graded S-module whose generators have squarefree degrees. Then all the shifts in the minimal $\mathbb{Z}^n$-graded free resolution of M are squarefree, and* $\operatorname{proj\,dim} M \leq n - \alpha(M)$.

In the squarefree case Theorem 4.33 can be improved as follows.

Theorem 4.36. *Let M be a $\mathbb{Z}^n$-graded S-module whose generators have squarefree degrees, and let $\mathbb{F}$ be a minimal $\mathbb{Z}^n$-graded free resolution of M with syzygy modules Z_p. Let $1 \leq p \leq n - \alpha(M)$ be an integer, and let $\mathcal{F} = e_1, \ldots, e_m$ be any homogeneous basis of F_{p-1}. Then $\operatorname{in}_{\mathcal{F}}(Z_p) = \bigoplus_{j=1}^m I_j e_j$, where the minimal set of monomial generators of each I_j is a squarefree monomial ideal in at most $n - \alpha(M) - p$ variables.*

Proof. Let

$$\mathbb{F}\colon\ 0 \to F_q \xrightarrow{\varphi_q} F_{q-1} \xrightarrow{\varphi_{q-1}} \cdots \xrightarrow{\varphi_2} F_1 \xrightarrow{\varphi_1} F_0 \xrightarrow{\epsilon} M \to 0$$

be the minimal $\mathbb{Z}^n$-graded free resolution of M. By Corollary 4.35 all shifts in the resolution are squarefree and $q \leq n - \alpha(M)$. In particular, Z_{p-1} is minimally generated by the elements $\varphi_p(e_1), \ldots, \varphi_p(e_m)$, and each of these elements has a squarefree degree. Here we have set $Z_{p-1} = M$ and $\varphi_{p-1} = \epsilon$, if $p = 1$. For each j we obtain an exact sequence

$$0 \longrightarrow Z_p \cap \bigoplus_{k=j}^m Se_k \longrightarrow \bigoplus_{k=j}^m Se_k \xrightarrow{\psi_j} N_j \longrightarrow 0,$$

where $N_j = \sum_{k=j}^m S\varphi_{p-1}(e_k)$ and where ψ_j is the restriction of φ_{p-1} to $\bigoplus_{k=j}^m Se_k$. Since N_j is generated by elements of squarefree degree, it follows from Proposition 4.34 that $Z_p \cap \bigoplus_{k=j}^m Se_k$ is minimally generated by elements of squarefree degree. Thus Lemma 4.32 implies that $I_j e_j$ is also minimally generated by elements of squarefree degree. In particular, I_j is a squarefree monomial ideal. Since the resolution $\mathbb{F}$ is minimal and e_j is a basis element of F_p, the vector $\deg e_j$ has at least $\alpha(M) + p$ nonzero entries. This implies that the degree of each generator of I_j can have at most $n - \alpha(M) - p$ nonzero entries. This yields the desired conclusion. □

Problems

Problem 4.1. Prove that an arbitrary intersection of submodules of an R-module M is again a submodule of M.

Problem 4.2. Let $N_1, \ldots, N_r$ be submodules of the R-module M. Show that the set of all elements of the form $m_1 + m_2 + \cdots + m_r$ with $m_i \in N_i$ is a submodule of M. (It is called the **sum** of the modules N_i and denoted $N_1 + N_2 + \cdots + N_r$.)

Problem 4.3. The sum of the submodules $N_1, \ldots, N_r$ of M is called **direct**, if whenever $m_1 + m_2 + \cdots + m_r = m_1' + m_2' + \cdots + m_r'$ with $m_i, m_i' \in N_i$, then $m_i = m_i'$ for $i = 1, \ldots, r$. The direct sum is denoted $N_1 \oplus N_2 \oplus \cdots \oplus N_r$. Show that $N_1 + N_2 + \cdots + N_r = N_1 \oplus N_2 \oplus \cdots \oplus N_r$ if and only if $N_i \cap (N_1 + \cdots + N_{i-1} + N_{i+1} + \cdots + N_r) = \{0\}$ for all i.

Problem 4.4. Let F be a free R-module with basis $e_1, \ldots, e_r$, and let Re_i be the submodule of F generated by e_i. Show that $F = Re_1 \oplus Re_2 \oplus \cdots \oplus Re_r$.

Problem 4.5. Let F be a free R-module with basis $e_1, \ldots, e_r$ and let $I \subset R$ be an ideal of R. Show that the elements $e_1 + IF, \ldots, e_r + IF$ form a basis of the R/I-module F/IF.

Problem 4.6. (a) Let R be a ring and $0 \to U \to M \to N \to 0$ a short exact sequence of R-modules, and let $x \in R$ be an element which is regular on M and N. Show that x is also regular on U, and that the induced sequence $0 \to U/xU \to M/xM \to N/xN \to 0$ is exact.

(b) Let M be a finitely generated graded S-module, $\mathbb{F}$ a graded free S-resolution of M and Z_p the pth syzygy module of M with respect to $\mathbb{F}$. Use (a) to show that if $f_1, \ldots, f_p$ is a homogeneous S-sequence, then it is a Z_p-sequence as well.

Problem 4.7. Let F be a free S-module with basis e_1, e_2. Find a monomial order on F which is not of the type "coefficient before position" or "position before coefficient", as described in Section 4.2.

Problem 4.8. Let $<$ be a monomial order on the free S-module F, and let $N \subset M \subset F$ be two submodules of F. Show that $N = M$ if and only if $\mathtt{in}_<(N) = \mathtt{in}_<(M)$.

Problem 4.9. In analogy to Theorem 2.6, prove the following: let $<$ be a monomial order on the free S-module F, and let $U \subset F$ be a submodule of F. Then the monomials not belonging to $\mathtt{in}_<(U)$ form a K-basis of F/U.

Problem 4.10. Compute, with respect to the lexicographic order, the Gröbner basis of the module given at the end of Section 4.3.

Problem 4.11. Let $<$ be a monomial order on the free S-module $F = \bigoplus_{j=1}^m Se_j$, let $U \subset F$ be a submodule of F, and suppose that $\mathtt{in}_<(U) = \bigoplus_{j=1}^m I_j e_j$. Show that U is a free S-module if and only if I_j is a principal ideal for $j = 1, \ldots, m$.

Problem 4.12. Let I be a monomial ideal generated by $u_1, \ldots, u_m$, and let $\epsilon\colon \bigoplus_{j=1}^n Se_j \to I$ be the epimorphism with $e_i \mapsto u_i$ for $i = 1, \ldots, m$. Show that $\operatorname{Ker} \epsilon$ is generated by the relations

$$\frac{\operatorname{lcm}(u_i, u_j)}{u_i} e_i - \frac{\operatorname{lcm}(u_i, u_j)}{u_j} e_j \quad \text{with} \quad 1 \leq i < j \leq m.$$

Problem 4.13. Show that the generators of $\operatorname{Ker} \epsilon$ as described in Problem 4.12 form a Gröbner basis with respect to any monomial order which gives priority to the position.

Problem 4.14. (Unpublished observation by Tom Sederberg) Fix polynomials $f_1, \ldots, f_m \in S$ and a monomial order $<$ on S. Let S^{m+1} have basis $g_0, \ldots, g_m$ and consider the submodule M of S^{m+1} generated by $f_i g_0 + g_i$ for $i = 1, ..., m$. Let $\mathcal{G}$ be a Gröbner basis of M with respect to the position over coefficient monomial order on S^{m+1} coming from $<$. Prove the following:

(i) The elements of $\mathcal{G}$ with zero first coordinate give a Gröbner basis of $\operatorname{Syz}(f_1, ..., f_m)$ with respect to the position over coefficient monomial order on S^m coming from $<$.

(ii) The nonzero first coordinates of elements of $\mathcal{G}$ give a Gröbner basis of $(f_1, ..., f_m)$ with respect to $<$. Furthermore, if $h_0 \in S$ is one of these elements, then the corresponding element $(h_0, h_1, ..., h_m)$ of $\mathcal{G}$ satisfies $h_0 = h_1 f_1 + \cdots + h_m f_m$. Thus we get explicit expressions for the elements of the Gröbner basis of the ideal $(f_1, \ldots, f_m)$ in terms of the original generators $f_1, ..., f_m$.

Problem 4.15. The ideal $I = (x^2 - yz, x^3 - w^3, y^2 - zw, 2xy + zw, xyzw) \subset S = \mathbb{Q}[x, y, z, w]$ has the resolution

$$0 \to S^4(-7) \to S^{12}(-6) \to S^3(-4) \oplus S^9(-5) \to S^3(-2) \oplus S(-3) \oplus S(-4) \to I.$$

Compute depth, dimension, regularity, multiplicity and a-invariant of S/I.

Problem 4.16. Let $I \subset S = K[x_1, \ldots, x_n]$ be a squarefree monomial ideal whose least degree of a generator is a. Show that $\operatorname{proj\,dim} I \leq n - a$.

Problem 4.17. As an extension of Corollary 4.35, prove the following result: Let M be a $\mathbb{Z}^n$-graded S-module generated by the homogeneous elements $m_1 \ldots, m_r$ with $\deg m_i = \mathbf{a}_i \in \mathbb{Z}^n_{\geq 0}$. Let $\mathbf{a} \in \mathbb{Z}^n$ be the vector which is componentwise the maximum of the vectors $\mathbf{a}_i$. Then all shifts in a minimal $\mathbb{Z}^n$-graded free resolution of M are componentwise $\leq \mathbf{a}$.

Chapter 5

Gröbner bases of toric ideals

Semigroup rings and their toric ideals are widely studied from an algebraic and combinatorial point of view and have applications in other fields like algebraic combinatorics, algebraic statistics, etc. This rich class of rings includes the polynomial ring $S = K[x_1, \ldots, x_n]$ which is associated with the semigroup $\mathbb{N}^n$. Given a semigroup ring one may study various properties of it such as normality, Cohen-Macaulayness, Koszulness. Gröbner basis theory plays a main role in this study. This chapter aims at giving a quick insight into this subject.

5.1. Semigroup rings and toric ideals

In this section we give the fundamental definitions concerning semigroup rings and their toric ideals and prove that a binomial ideal is toric if and only if it is prime.

Let $\mathbb{Z}$ be the set of integers and $\{\mathbf{h}_1, \ldots, \mathbf{h}_q\}$ a subset of $\mathbb{Z}^n$ where n is a positive integer. Let H be the submonoid of the additive group $\mathbb{Z}^n$ generated by $\mathbf{h}_1, \ldots, \mathbf{h}_q$, that is,

$$H = \mathbb{N}\mathbf{h}_1 + \cdots + \mathbb{N}\mathbf{h}_q = \{a_1\mathbf{h}_1 + \cdots + a_q\mathbf{h}_q : a_i \in \mathbb{N} \text{ for } 1 \leq i \leq q\},$$

where $\mathbb{N}$ is the set of nonnegative integers. H is called an **affine semigroup**.

Let $K[x_1, x_1^{-1}, \ldots, x_n, x_n^{-1}]$ be the Laurent polynomial ring and $K[H]$ the subring of $K[x_1, x_1^{-1}, \ldots, x_n, x_n^{-1}]$ generated over K by the monomials $f_i = \mathbf{x}^{\mathbf{h}_i}$ for $i = 1, \ldots, q$, where $\mathbf{x}^{\mathbf{c}} = x_1^{c_1} \cdots x_n^{c_n}$ if $\mathbf{c} = (c_1, \ldots, c_n) \in \mathbb{Z}^n$.

Obviously, $\mathbf{x}^{\mathbf{g}} \in K[H]$ if and only if $\mathbf{g} \in H$. The elements of $K[H]$ are polynomial expressions in $f_1, \ldots, f_q$ with coefficients in K, that is, of the form $\sum_{\mathbf{a}} c_{\mathbf{a}} \mathbf{f}^{\mathbf{a}}$ with $c_{\mathbf{a}} \in K$, where $\mathbf{f}^{\mathbf{a}} = f_1^{a_1} \cdots f_q^{a_q}$ if $\mathbf{a} = (a_1, \ldots, a_q)$. The K-algebra $K[H]$ is called the **semigroup ring** associated with the affine semigroup H. For example, $K[\mathbb{Z}^n] = K[x_1, x_1^{-1}, \ldots, x_n, x_n^{-1}]$ and $K[\mathbb{N}^n] = K[x_1, \ldots, x_n]$.

In the sequel we consider affine semigroups which are generated by finitely many vectors $\mathbf{h}_1, \ldots, \mathbf{h}_q \in \mathbb{N}^n$. In this case, $K[H]$ is a K-subalgebra of the polynomial ring $S = K[x_1, \ldots, x_n]$. Let $R = K[t_1, \ldots, t_q]$ be the polynomial ring in the indeterminates $t_1, \ldots, t_q$ and $\varphi : R \to K[H]$ the K-algebra homomorphism defined by $t_i \mapsto f_i = \mathbf{x}^{\mathbf{h}_i}$ for $i = 1, \ldots, q$.

One may consider different gradings on $K[H]$. For instance, $K[H]$ is $\mathbb{Z}$-graded with the $\mathbb{Z}$-grading induced from S, that is, $K[H]_d = K[H] \cap S_d$, where S_d is the degree d homogeneous component of S. In this case, $\varphi : R \to K[H]$ is a graded homomorphism if we set $\deg(t_i) = \deg(f_i)$ for $i = 1, \ldots, q$. When all the generators of $K[H]$ have the same degree, let us say d, then $K[H]$ may be even viewed as a standard graded algebra if we set $K[H]_i = K[H] \cap S_{di}$ for $i \geq 0$. Obviously, in this case, $\varphi : R \to K[H]$ is a graded homomorphism of standard graded K-algebras.

On the other hand, one may consider the natural $\mathbb{Z}^n$-grading on S and the induced grading on $K[H]$, that is, $K[H]_{\mathbf{a}} = K[H] \cap S_{\mathbf{a}}$ for all $\mathbf{a} \in \mathbb{Z}^n$. With respect to this grading of $K[H]$, $\varphi : R \to K[H]$ is again graded if we assign to each t_i the $\mathbb{Z}^n$-degree of f_i.

φ is clearly surjective. Its kernel, denoted P_H, is a graded ideal of R. Moreover, since $K[H] \cong R/P_H$ and $K[H] \subset S$ is a domain, it follows that P_H is a prime ideal. P_H is called the **toric ideal** of H.

For a polynomial $h \in R$, we have $\varphi(h) = h(f_1, \ldots, f_q)$. In particular, for a monomial $\mathbf{t}^{\mathbf{u}} = t_1^{u_1} \cdots t_q^{u_q} \in R$, $\varphi(\mathbf{t}^{\mathbf{u}}) = f_1^{u_1} \cdots f_q^{u_q} = \mathbf{x}^{\sum_{i=1}^q u_i \mathbf{h}_i}$. Let $\pi : \mathbb{N}^q \to H$ be the homomorphism of semigroups defined by $\pi(\mathbf{u}) = \sum_{i=1}^q u_i \mathbf{h}_i$ for $\mathbf{u} = (u_1, \ldots, u_q) \in \mathbb{N}^q$. Then $\varphi(\mathbf{t}^{\mathbf{u}}) = \mathbf{x}^{\pi(\mathbf{u})}$ for $\mathbf{t}^{\mathbf{u}} \in R$.

Notice that we already have an algorithm to compute a Gröbner basis of the toric ideal P_H which follows from Corollary 3.9. Namely, if we consider the ideal $J = (t_1 - f_1, \ldots, t_q - f_q) \subset K[t_1, \ldots, t_q, x_1, \ldots, x_n]$, then $P_H = J \cap K[t_1, \ldots, t_q]$. Therefore, if $\mathcal{G}$ is a Gröbner basis of J with respect to an elimination order for $x_1, \ldots, x_n$, then the set $\mathcal{G} \cap R$ is a Gröbner basis of P_H.

Let us consider a few simple examples.

Example 5.1. (i) Let $H \subset \mathbb{N}$ be the semigroup generated by $3, 4, 5$. Then $K[H] = K[x^3, x^4, x^5] \subset K[x]$. We have $P_H = J \cap K[t_1, t_2, t_3]$ where $J = (t_1 - x^3, t_2 - x^4, t_3 - x^5)$. The reduced Gröbner basis of J with respect to the lexicographic order induced by $x > t_1 > t_2 > t_3$ is $\{x^3 - t_1, xt_1 - t_2, xt_2 -$

$t_3, xt_3-t_1^2, t_1^3-t_2t_3, t_1^2t_2-t_3^2, t_1t_2^3-t_3^3, t_1t_3-t_2^2, t_2^5-t_3^4\}$. Therefore, the reduced Gröbner basis of P_H is $\{t_1^3 - t_2t_3, t_1^2t_2 - t_3^2, t_1t_2^3 - t_3^3, t_1t_3 - t_2^2, t_2^5 - t_3^4\}$.

(ii) Let $K[H] = K[x_1^2, x_1x_2, x_2^2] \subset K[x_1, x_2]$. The reduced Gröbner basis of $P_H \subset K[t_1, t_2, t_3]$ with respect to the lexicographic order is $\{t_1t_3 - t_2^2\}$.

(iii) Let $K[H] = K[x_1^4, x_1^3x_2, x_1x_2^3, x_2^4] \subset K[x_1, x_2]$. The reduced Gröbner basis of $P_H \subset K[t_1, \ldots, t_4]$ with respect to the lexicographic order is $\{t_1^2t_3 - t_2^3, t_1t_3^2 - t_2^2t_4, t_1t_4 - t_2t_3, t_2t_4^2 - t_3^3\}$.

We notice in the above examples that all the elements of the reduced Gröbner basis are binomials. We shall see in Proposition 5.6 that this is indeed the case for the reduced Gröbner basis of every toric ideal.

A **binomial** is a polynomial which is a difference of two monomials. A polynomial ideal is a **binomial ideal** if it is generated by a set of binomials. Note that a binomial ideal does not contain any monomial. Sometimes it is convenient to consider binomial ideals in a more general form; see for instance, [**ES84**], [**Mi10**].

The following lemma shows that any toric ideal can be generated by binomials.

Lemma 5.2. *Let P_H be the toric ideal of the affine semigroup ring H. Then the set of binomials $\{\mathbf{t}^{\mathbf{u}} - \mathbf{t}^{\mathbf{v}} : \mathbf{u}, \mathbf{v} \in \mathbb{N}^q, \pi(\mathbf{u}) = \pi(\mathbf{v})\}$ generates P_H as a K-vector space. In particular, P_H is a binomial ideal.*

Proof. Obviously $\mathbf{t}^{\mathbf{u}} - \mathbf{t}^{\mathbf{v}} \in P_H$ for all $\mathbf{u}, \mathbf{v} \in \mathbb{N}^q$ such that $\pi(\mathbf{u}) = \pi(\mathbf{v})$. On the other hand, since P_H is $\mathbb{Z}^n$-graded, it is enough to look at the polynomials of P_H which are homogeneous with respect to the $\mathbb{Z}^n$-grading. Let $f = c_1\mathbf{t}^{\mathbf{u}_1} + \cdots + c_s\mathbf{t}^{\mathbf{u}_s}$ be a homogeneous polynomial in P_H, where $c_1, \ldots, c_s \in K \setminus \{0\}$. By the assumption on f, it follows that all the monomials $\mathbf{t}^{\mathbf{u}_i}$ are mapped to the same monomial $\mathbf{x}^{\mathbf{h}}$ under φ, where $\mathbf{h} = \pi(\mathbf{u}_1) = \cdots = \pi(\mathbf{u}_s)$. Then we get $0 = \varphi(f) = (\sum_{i=1}^s c_i)\mathbf{x}^{\mathbf{h}}$, whence $\sum_{i=1}^s c_i = 0$. We obtain $c_1 = -c_2 - \cdots - c_s$. Consequently, we can write $f = c_2(\mathbf{t}^{\mathbf{u}_2} - \mathbf{t}^{\mathbf{u}_1}) + \cdots + c_s(\mathbf{t}^{\mathbf{u}_s} - \mathbf{t}^{\mathbf{u}_1})$, thus the proof is complete. □

As we have already observed, every toric ideal is a binomial prime ideal. We show in the sequel that if a binomial ideal is prime, then it is a toric ideal.

For a vector $\mathbf{v} \in \mathbb{Z}^q$, we denote by $\mathbf{v}^+$ and $\mathbf{v}^-$ the vectors with nonnegative components defined as follows:

$$v_i^+ = \begin{cases} v_i, & \text{if } v_i \geq 0, \\ 0, & \text{if } v_i < 0, \end{cases} \quad \text{and } v_i^- = \begin{cases} 0, & \text{if } v_i > 0, \\ -v_i, & \text{if } v_i \leq 0. \end{cases}$$

Obviously, we have $\mathbf{v} = \mathbf{v}^+ - \mathbf{v}^-$, and this decomposition of $\mathbf{v}$ into vectors with nonnegative coefficients is unique. In particular, every binomial in R

whose monomials have disjoint supports can be written as $f_{\mathbf{v}} = \mathbf{t}^{\mathbf{v}^+} - \mathbf{t}^{\mathbf{v}^-}$ where $\mathbf{v}$ is a vector in $\mathbb{Z}^q$.

One easily checks that for any $\mathbf{v}, \mathbf{w} \in \mathbb{Z}^q$ we have

$$f_{\mathbf{v}} f_{\mathbf{w}} = u f_{\mathbf{v}+\mathbf{w}} - \mathbf{t}^{\mathbf{v}^-} f_{\mathbf{w}} - \mathbf{t}^{\mathbf{w}^-} f_{\mathbf{v}}, \tag{5.1}$$

for some monomial $u \in R$. Indeed, the identity follows straightforwardly if we observe that

$$\mathbf{v}^+ + \mathbf{w}^+ - (\mathbf{v}+\mathbf{w})^+ = \mathbf{v}^- + \mathbf{w}^- - (\mathbf{v}+\mathbf{w})^-$$

for any two vectors $\mathbf{v}, \mathbf{w} \in \mathbb{Z}^q$.

A **lattice** is a submodule of a free $\mathbb{Z}$-module of finite rank. With each lattice one may associate a binomial ideal.

Definition 5.3. *Let $L \subset \mathbb{Z}^q$ be a lattice. The* ***lattice ideal*** *associated with L is the binomial ideal $I_L = (\mathbf{t}^{\mathbf{v}^+} - \mathbf{t}^{\mathbf{v}^-} : \mathbf{v} \in L)$.*

Proposition 5.4. *Let $I \subset R$ be a binomial prime ideal. Then I is a lattice ideal.*

Proof. Since I is a prime ideal we may assume that its generators are of the form $f_{\mathbf{v}}$ with $\mathbf{v} \in \mathbb{Z}^q$. It is obvious that if $f_{\mathbf{v}} = \mathbf{t}^{\mathbf{v}^+} - \mathbf{t}^{\mathbf{v}^-} \in I$, then $f_{-\mathbf{v}} = -f_{\mathbf{v}} \in I$. Therefore, it is enough to show that if $f_{\mathbf{v}}, f_{\mathbf{w}} \in I$, then $f_{\mathbf{v}+\mathbf{w}} \in I$. Let $f_{\mathbf{v}}, f_{\mathbf{w}} \in I$. By using the identity (5.1), we get that there exists a monomial $u \in R$ such that $u f_{\mathbf{v}+\mathbf{w}} \in I$. But, obviously, $u \notin I$, thus $f_{\mathbf{v}+\mathbf{w}} \in I$ since I is a prime ideal. □

We are ready to prove the following

Theorem 5.5. *Let $I \subset R$ be a binomial prime ideal. Then I is a toric ideal.*

Proof. By Proposition 5.4, $I = I_L$ where $L \subset \mathbb{Z}^q$ is the lattice generated by the vectors $\mathbf{v} \in \mathbb{Z}^q$ with the property that the associated binomials $f_{\mathbf{v}}$ generate I.

We first show that the factor module $\mathbb{Z}^q/L$ is torsion free, thus it is free. In other words, we have to show that if $\mathbf{v} \in \mathbb{Z}^q$ and $m > 1$ is an integer such that $m\mathbf{v} \in L$, equivalently, $f_{m\mathbf{v}} \in I$, then $\mathbf{v} \in L$, that is, $f_{\mathbf{v}} \in I$.

Let $f_{m\mathbf{v}} = \mathbf{t}^{m\mathbf{v}^+} - \mathbf{t}^{m\mathbf{v}^-} \in L$. If $\mathrm{char}(K) = 0$, then we decompose $f_{m\mathbf{v}} = f_{\mathbf{v}} g$ where $g = \mathbf{t}^{(m-1)\mathbf{v}^+} + \mathbf{t}^{(m-2)\mathbf{v}^+}\mathbf{t}^{\mathbf{v}^-} + \cdots + \mathbf{t}^{\mathbf{v}^+}\mathbf{t}^{(m-2)\mathbf{v}^-} + \mathbf{t}^{(m-1)\mathbf{v}^-} \in R$. By using the substitutions $t_i \mapsto 1$ for $i = 1, \ldots, q$, we easily see that $g \notin I$ since all binomials vanish on this substitution. Therefore, $f_{\mathbf{v}} \in I$ since I is a prime ideal.

If $\mathrm{char}(K) = p > 0$, then we write $m = p^e m'$ where $e \geq 0$, $m' \geq 1$ are integers such that p does not divide m'. In this case we decompose $f_{m\mathbf{v}}$ as

$f_{\mathbf{v}}^{p^e} g'$ where $g' = (\mathbf{t}^{p^e\mathbf{v}^+})^{m'-1} + \cdots + (\mathbf{t}^{p^e\mathbf{v}^-})^{m'-1} \in R$. By using again the substitutions $t_i \mapsto 1$ for $i = 1, \ldots, q$, we get $g' \notin I$. Since I is a prime ideal, it follows that $f_{\mathbf{v}}^{p^e} \in I$, whence $f_{\mathbf{v}} \in I$.

Let n be the rank of the free module $\mathbb{Z}^q/L$ and $\bar{\mathbf{h}}_1, \ldots, \bar{\mathbf{h}}_q$ the generators of $\mathbb{Z}^q/L$ viewed as vectors with n components. Let $K[H] \subset K[x_1^{\pm 1}, \ldots, x_n^{\pm 1}]$ be the semigroup ring generated by $\mathbf{x}^{\bar{\mathbf{h}}_1}, \ldots, \mathbf{x}^{\bar{\mathbf{h}}_q}$ and $\varphi : R \to K[H]$ the K-algebra homomorphism defined by $t_i \mapsto \mathbf{x}^{\bar{\mathbf{h}}_i}$ for $i = 1, \ldots, q$. Then, by using Problem 5.6, one easily shows that the kernel of φ is exactly I, thus I is a toric ideal. □

5.2. Gröbner bases of toric ideals

In this section we give basic properties of Gröbner bases of toric ideals.

Proposition 5.6. *Let H be an affine semigroup and P_H its toric ideal. Let $<$ be a monomial order on R and $\mathcal{G}$ the reduced Gröbner basis of P_H with respect to $<$. Then, every element of $\mathcal{G}$ is a binomial of the form $\mathbf{t}^{\mathbf{u}} - \mathbf{t}^{\mathbf{v}}$ with $\pi(\mathbf{u}) = \pi(\mathbf{v})$.*

Proof. Let G be a set of generators of P_H. By Lemma 5.2, we can choose these generators of the form $\mathbf{t}^{\mathbf{u}} - \mathbf{t}^{\mathbf{v}}$ with $\pi(\mathbf{u}) = \pi(\mathbf{v})$. By applying the Buchberger algorithm, at each step when we get a nonzero remainder, this is again a binomial, let us say $\mathbf{t}^{\mathbf{w}} - \mathbf{t}^{\mathbf{w}'}$, with $\pi(\mathbf{w}) = \pi(\mathbf{w}')$. Therefore, P_H has a Gröbner basis with respect to the given order formed with binomials of the form $\mathbf{t}^{\mathbf{u}} - \mathbf{t}^{\mathbf{v}}$ with $\pi(\mathbf{u}) = \pi(\mathbf{v})$. By applying the reductions needed for getting the reduced Gröbner basis, we get again binomials of this type. □

The above proposition shows that any reduced Gröbner basis of P_H consists of monomials $\mathbf{t}^{\mathbf{u}} - \mathbf{t}^{\mathbf{v}}$ with $\pi(\mathbf{u}) = \pi(\mathbf{v})$. A nice property of the binomials of any reduced Gröbner basis of a toric ideal is given below.

Definition 5.7. *A binomial $\mathbf{t}^{\mathbf{u}} - \mathbf{t}^{\mathbf{v}} \in P_H$ is called **primitive** if there is no other binomial $\mathbf{t}^{\mathbf{r}} - \mathbf{t}^{\mathbf{s}} \in P_H$ such that $\mathbf{t}^{\mathbf{r}} | \mathbf{t}^{\mathbf{u}}$ and $\mathbf{t}^{\mathbf{s}} | \mathbf{t}^{\mathbf{v}}$.*

Note that if $\mathbf{t}^{\mathbf{u}} - \mathbf{t}^{\mathbf{v}}$ is a primitive binomial, then the monomials $\mathbf{t}^{\mathbf{u}}$ and $\mathbf{t}^{\mathbf{v}}$ have disjoint supports.

Proposition 5.8. *Let P_H be the toric ideal of the affine semigroup H and $\mathcal{G}$ the reduced Gröbner basis of P_H with respect to a monomial order. Then any binomial of $\mathcal{G}$ is primitive.*

Proof. Let $g = \mathbf{t}^{\mathbf{u}} - \mathbf{t}^{\mathbf{v}}$ be a binomial of $\mathcal{G}$ with $\mathrm{in}_<(g) = \mathbf{t}^{\mathbf{u}}$. Since $\mathcal{G}$ is reduced, it follows that $\mathbf{t}^{\mathbf{v}} \notin \mathrm{in}_<(P_H)$. Let us assume that there exists $h = \mathbf{t}^{\mathbf{r}} - \mathbf{t}^{\mathbf{s}} \in P_H$, $h \neq g$, such that $\mathbf{t}^{\mathbf{r}} | \mathbf{t}^{\mathbf{u}}$ and $\mathbf{t}^{\mathbf{s}} | \mathbf{t}^{\mathbf{v}}$. If $\mathrm{in}_<(h) = \mathbf{t}^{\mathbf{r}}$, then we must have $\mathbf{t}^{\mathbf{r}} = \mathbf{t}^{\mathbf{u}}$. It follows that $h' = \mathbf{t}^{\mathbf{v}} - \mathbf{t}^{\mathbf{s}}$ belongs to P_H and

$\mathtt{in}_<(h') = \mathbf{t}^{\mathbf{v}}$ since $\mathbf{t}^{\mathbf{s}}|\mathbf{t}^{\mathbf{v}}$, a contradiction. Therefore, $\mathtt{in}_<(h) = \mathbf{t}^{\mathbf{s}}$, which leads again to contradiction since $\mathbf{t}^{\mathbf{s}}|\mathbf{t}^{\mathbf{v}}$. □

5.3. Simplicial complexes and squarefree monomial ideals

A basic construction which relates the simplicial complexes to commutative algebra consists in associating with any abstract simplicial complex Δ a squarefree monomial ideal and the quotient ring modulo this ideal. The corresponding ideal (ring) is called the **Stanley-Reisner ideal (ring)** of Δ. This construction has become a fundamental tool in combinatorial commutative algebra due to the work of Stanley, Hochster and Reisner ([**Ho77**], [**S75**], [**S96**], [**R76**]).

Here we briefly review this correspondence and some of its properties which will be useful later.

For a positive integer n we denote $[n] = \{1, \ldots, n\}$. A **simplicial complex** Δ on the set $[n]$ is a collection of subsets of $[n]$ which is closed under taking subsets, that is, for every $F \in \Delta$, if $G \subset F$, then $G \in \Delta$. We recall the standard terminology in this frame. Every element $F \in \Delta$ is called a **face** of Δ. The **dimension** of $F \in \Delta$ is $\dim F = |F| - 1$. The empty set is always a face of Δ unless Δ itself is the empty complex. $\emptyset$ is the unique face of Δ of dimension -1. The maximal faces with respect to inclusion are called **facets**. If $F_1, \ldots, F_m$ are the facets of Δ, we write $\Delta = \langle F_1, \ldots, F_m \rangle$ and say that Δ is **generated** by $F_1, \ldots, F_m$. We will also denote the set of facets of Δ by $\mathcal{F}(\Delta)$. The **dimension** of Δ is $\dim \Delta = \max\{\dim F : F \in \mathcal{F}(\Delta)\}$. A simplicial complex whose facets have the same dimension is called **pure**.

Let $f_i = f_i(\Delta)$ be the number of the faces of Δ of dimension i. If $\dim \Delta = d - 1$, we denote $f = f(\Delta) = (f_{-1}, f_0, \ldots, f_{d-1})$ and call this vector the **f-vector** of Δ. For example, the **n-simplex** is the simplicial complex which has a unique facet, namely $F = [n]$. Its f-vector is $f = (1, \binom{n}{1}, \ldots, \binom{n}{n-1}, 1)$. The boundary of the n-simplex is the simplicial complex with the facets $F_i = [n] \setminus \{i\}$ where $i = 1, \ldots, n$. Obviously, its f-vector is $(1, \binom{n}{1}, \ldots, \binom{n}{n-1})$. The simplicial complex which consists of n isolated vertices, that is, whose facets are $\{1\}, \ldots, \{n\}$, has the f-vector $(1, n)$.

Let $S = K[x_1, \ldots, x_n]$ be the polynomial ring. For a subset $F \subset [n]$ we denote by $\mathbf{x}_F$ the squarefree monomial whose support is F, that is, $\mathbf{x}_F = \prod_{i \in F} x_i$. With each simplicial complex Δ one may associate a squarefree monomial ideal $I_\Delta \subset S$ which is generated by all the squarefree monomials $\mathbf{x}_F$ with $F \notin \Delta$. Obviously, if $F \notin \Delta$, that is, F is a **nonface** of Δ, then $G \notin \Delta$ for any $G \subset [n]$ such that $G \supset F$. Therefore, I_Δ is minimally generated by the minimal nonfaces of Δ with respect to inclusion. I_Δ is the **Stanley-Reisner ideal** of Δ. The quotient ring $K[\Delta] = S/I_\Delta$ is the **Stanley-Reisner ring** of Δ.

For example, let Δ be the simplicial complex on the set $[5]$ with the facets $\{1,2,3\},\{2,4\},\{3,4\},\{4,5\}$. Its minimal nonfaces are $\{1,4\}$, $\{1,5\}$, $\{2,5\}$, $\{3,5\}$ and $\{2,3,4\}$. Consequently, $I_\Delta = (x_1x_4, x_1x_5, x_2x_5, x_3x_5, x_2x_3x_4)$. The Stanley-Reisner ideal associated to the n-simplex is the zero ideal. If Δ is the boundary of the n-simplex, then $I_\Delta = (x_1 \cdots x_n)$. The simplicial complex which is given by n isolated vertices has the Stanley-Reisner ideal $I_\Delta = (x_ix_j : 1 \le i < j \le n)$.

The correspondence *simplicial complexes* $\mapsto$ *squarefree monomial ideals* establishes a bijection between the simplicial complexes on the set $[n]$ and the squarefree monomial ideals in S. Indeed, if $I \subset S$ is a squarefree monomial ideal, one may consider the collection $\Delta = \{F \subset [n] : \mathbf{x}_F \notin I\}$. It is easily seen that Δ is indeed a simplicial complex and that $I = I_\Delta$.

Let Δ be a simplicial complex on the set $[n]$ of dimension $d-1$. As we have seen in Subsection 4.4.6, the Hilbert series of $K[\Delta]$, $\mathrm{Hilb}_{K[\Delta]}(t)$, encodes the Krull dimension, $\dim K[\Delta]$, of the Stanley-Reisner ring. This dimension is closely related to the dimension of Δ.

Theorem 5.9. *Let Δ be a simplicial complex on the set $[n]$ with the f-vector $f = (f_{-1}, f_0, \dots, f_{d-1})$. Then*

$$\mathrm{Hilb}_{K[\Delta]}(t) = \frac{\sum_{i=0}^{d} f_{i-1}t^i(1-t)^{d-i}}{(1-t)^d}.$$

In particular, $\dim K[\Delta] = \dim \Delta + 1$.

Proof. One easily observes, by using Macaulay's Theorem (Theorem 2.6), that $K[\Delta]$ can be decomposed as follows as a K-vector space,

$$K[\Delta] = \bigoplus_{F \in \Delta} \mathbf{x}_F K[\{x_i : i \in F\}].$$

Therefore,

$$\mathrm{Hilb}_{K[\Delta]}(t) = \sum_{F \in \Delta} \frac{t^{|F|}}{(1-t)^{|F|}} = \sum_{i=0}^{d} \frac{f_{i-1}t^i}{(1-t)^i},$$

which gives the desired formula. □

Let Δ be a simplicial complex on the set $[n]$ and I_Δ its Stanley-Reisner ideal. Since I_Δ is a squarefree monomial ideal, by using Proposition 1.15, it follows that I_Δ is a finite intersection of monomial prime ideals. Moreover, the irredundant presentation of I_Δ as an intersection of monomial prime ideals can be given.

Proposition 5.10. *Let Δ be a simplicial complex on the set $[n]$. Then*

$$I_\Delta = \bigcap_{F \in \mathcal{F}(\Delta)} P_{F^c}$$

where $P_{F^c} = (x_i : i \in [n]\setminus F)$ *for* $F \in \mathcal{F}(\Delta)$. *In particular,* $\{P_{F^c} : F \in \mathcal{F}(\Delta)\}$ *is the set of all the minimal primes of* I_Δ.

Proof. Let $\mathbf{x}_G$ be a squarefree monomial in S. Then $\mathbf{x}_G \in \bigcap_{F\in\mathcal{F}(\Delta)} P_{F^c}$ if and only if, for every $F \in \mathcal{F}(\Delta)$, we have $G \cap F^c \neq \emptyset$, which is equivalent to saying that $G \not\subset F$ for every $F \in \mathcal{F}(\Delta)$. Therefore, $\mathbf{x}_G \in \bigcap_{F\in\mathcal{F}(\Delta)} P_{F^c}$ if and only if $G \notin \Delta$, that is, $\mathbf{x}_G \in I_\Delta$. □

From the above proposition, we get

$$\dim K[\Delta] = \max\{\dim(S/P) : P \text{ is a minimal prime ideal of } I_\Delta\}.$$

The inequality $\operatorname{depth} M \leq \dim(S/P)$ for all $P \in \operatorname{Ass}(M)$ and for every graded S-module M [**BH98**, Proposition 1.2.13.] is well known. In particular, it follows that $\operatorname{depth} M \leq \dim M$ for every graded S-module M, as we have already noticed in Subsection 4.4.6. Recall that the module M is Cohen-Macaulay if $\operatorname{depth} M = \dim M$.

Definition 5.11. *A simplicial complex* Δ *on the set* $[n]$ *is called* ***Cohen-Macaulay*** *over the field* K *if* $K[\Delta]$ *is Cohen-Macaulay.*

An immediate consequence of the above definition is that any Cohen-Macaulay complex is pure.

One of the most useful tools for proving that a simplicial complex is Cohen-Macaulay is shellability.

Definition 5.12. *Let* Δ *be a pure simplicial complex.* Δ *is called* ***shellable*** *if its facets can be ordered as* $F_1, \ldots, F_m$ *such that* $\langle F_1, \ldots, F_{i-1}\rangle \cap \langle F_i\rangle$ *is generated by a nonempty set of maximal proper faces of* F_i *for* $i = 2, \ldots, m$.

For example, let Δ be the boundary of the n-simplex whose facets are $F_i = [n]\setminus\{i\}$ for $i = 1, \ldots, n$. Then, for all $i = 2, \ldots, n$, $\langle F_1, \ldots, F_{i-1}\rangle \cap \langle F_i\rangle$ is generated by the maximal proper faces $F_i \setminus \{j\}$, $j = 1, \ldots, i-1$, of F_i. Therefore, it follows that Δ is a shellable complex.

Theorem 5.13. *Let* Δ *be a shellable simplicial complex. Then* Δ *is Cohen-Macaulay over any field.*

Proof. Let K be a field and assume that Δ has dimension $d-1$ and it is shellable with respect to the order $F_1, \ldots, F_m$ of its facets.. We apply induction on the number m of facets of Δ. For $m = 1$ there is nothing to prove. Let $m > 1$ and $J = \bigcap_{i=1}^{m-1} P_{F_i^c}$. Then $I_\Delta = J \cap P_{F_m^c}$. We have the exact sequence of graded S-modules (see Problem 5.9),

$$0 \to \frac{S}{I_\Delta} \to \frac{S}{J} \oplus \frac{S}{P_{F_m^c}} \to \frac{S}{J + P_{F_m^c}} \to 0.$$

Let Γ be the simplicial complex generated by $F_1, \dots, F_{m-1}$ on the vertex set $V' = \bigcup_{i=1}^{m-1} F_i$. We may assume that $V' = \{x_1, \dots, x_r\}$ for some $1 \leq r \leq n$. Then $J = I_\Gamma + (x_{r+1}, \dots, x_n)$ and $S/J \cong K[x_1, \dots, x_r]/I_\Gamma$ is Cohen-Macaulay of dimension d by induction. We also have $S/P_{F_m^c} \cong K[\{x_i : i \in F_m\}]$, thus $S/P_{F_m^c}$ is Cohen-Macaulay of dimension d. This shows that the middle term in the above exact sequence is Cohen-Macaulay of dimension d. If we show that the right term is Cohen-Macaulay of dimension $d-1$, it follows that $K[\Delta]$ is Cohen-Macaulay of dimension d. Indeed, in order to prove this claim one may use a standard argument which involves the Depth Lemma. We refer the reader to [**BH98**] or [**E95**] for more details. But

$$J + P_{F_m^c} = \bigcap_{i=1}^{m-1} P_{F_i^c} + P_{F_m^c}$$

$$= \bigcap_{i=1}^{m-1} (P_{F_i^c} + P_{F_m^c}) = \bigcap_{i=1}^{m-1} (x_j : j \in F_i^c \cup F_m^c) = P_{F_m^c} + (w),$$

where $w = \prod_{t \in \bigcup_{i=1}^{m-1} (F_m \setminus F_i)} x_t$. This gives

$$S/(J + P_{F_m^c}) \cong K[\{x_j : j \in F_m\}]/(w),$$

which shows that $S/(J + P_{F_m^c})$ is Cohen-Macaulay of dimension $d-1$. □

5.4. Normal semigroup rings

Let R be a domain with field of fractions $Q(R)$. The **normalization** or **integral closure** of R is the subring $\bar{R} \subset Q(R)$ which consists of all the elements of $Q(R)$ which are integral over R. The domain R is called **normal** if $R = \bar{R}$. In this section we study normal semigroup rings. The reader who is not familiar with the concept of normality may consult [**M86**] or [**E95**].

Let H be an arbitrary affine semigroup generated by $\mathcal{H} = \{\mathbf{h}_1, \dots, \mathbf{h}_q\} \subset \mathbb{Z}^n$. Let $\mathbb{Z}H$ be the subgroup of $\mathbb{Z}^n$ generated by $\mathcal{H}$. H is called **normal** if it satisfies the following condition: if $m\mathbf{g} \in H$ for some $\mathbf{g} \in \mathbb{Z}H$ and $m > 0$, then $\mathbf{g} \in H$.

It is easily seen that if the semigroup ring $K[H]$ is normal, then the semigroup H is normal as well. Indeed, let $K[H]$ be a normal ring and let $m\mathbf{g} \in H$ for some $\mathbf{g} \in \mathbb{Z}H$ and $m > 0$. Then $\mathbf{x}^{\mathbf{g}}$ belongs to the field of fractions of $K[H]$ and $\mathbf{x}^{m\mathbf{g}} \in K[H]$. Hence, if $\mathbf{x}^{m\mathbf{g}} = u$, then $\mathbf{x}^{\mathbf{g}}$ satisfies the integral equation $y^m - u$ over $K[H]$. By the normality of $K[H]$, we have $\mathbf{x}^{\mathbf{g}} \in K[H]$, which is equivalent to $\mathbf{g} \in H$.

The converse is also true. In order to prove it, we briefly recall some useful concepts.

Let $\mathcal{F} = \{\mathbf{v}_1, \ldots, \mathbf{v}_m\}$ be a finite subset of $\mathbb{R}^n$ and let $\mathbb{R}_+$ be the set of nonnegative real numbers. The set

$$\mathbb{R}_+\mathcal{F} = \{\sum_{i=1}^{m} a_i\mathbf{v}_i : a_i \in \mathbb{R}_+ \text{ for } 1 \leq i \leq m\}$$

is called the **cone** generated by $\mathcal{F}$.

For two vectors $\mathbf{c} = (c_1, \ldots, c_n), \mathbf{x} = (x_1, \ldots, x_n) \in \mathbb{R}^n$, we denoted by $\langle \mathbf{c}, \mathbf{x} \rangle$ the standard scalar product which is defined as $\langle \mathbf{c}, \mathbf{x} \rangle = c_1x_1 + \cdots c_nx_n$.

It is known [**BG09**, Theorem 1.15] that every finitely generated cone is the intersection of finitely many half-spaces $H_1^+, \ldots, H_m^+$ where

$$H_i^+ = \{\mathbf{x} \in \mathbb{R}^n : \langle \mathbf{c}_i, \mathbf{x} \rangle \geq 0\} \text{ for some } \mathbf{c}_i \in \mathbb{R}^n, \mathbf{c}_i \neq 0, i = 1, \ldots, m.$$

If all the half-spaces $H_1^+, \ldots, H_m^+$ are rational, that is, $\mathbf{c}_i \in \mathbb{Q}^n$ for all i, then the cone $H_1^+ \cap \ldots \cap H_m^+$ is called **rational**. When the cone is rational, one may choose the $\mathbf{c}_i$ defining the cone to be integral vectors. Any finitely generated rational cone is of the form $\mathbb{R}_+\mathcal{F}$, where $\mathcal{F}$ is a finite subset of $\mathbb{Q}^n$. One may also prove [**BG09**, Chapter 1] that if $\mathcal{F} = \{\mathbf{v}_1, \ldots, \mathbf{v}_m\} \subset \mathbb{Z}^n$, then

$$\mathbb{Z}\mathcal{F} \cap \mathbb{R}_+\mathcal{F} = \mathbb{Z}\mathcal{F} \cap \mathbb{Q}_+\mathcal{F}, \tag{5.2}$$

where $\mathbb{Z}\mathcal{F}$ is the subgroup of $\mathbb{Z}^n$ generated by $\mathcal{F}$ and $\mathbb{Q}_+\mathcal{F} = \{\sum_{i=1}^m a_i\mathbf{v}_i : a_i \in \mathbb{Q}_+\}$. Here we denoted by $\mathbb{Q}_+$ the set of nonnegative rational numbers. Moreover,

$$\mathbb{Z}^n \cap \mathbb{R}_+\mathcal{F} = \mathbb{Z}^n \cap \mathbb{Q}_+\mathcal{F}. \tag{5.3}$$

The next proposition is known as **Gordan's Lemma**.

Proposition 5.14. (a) *If H is a normal affine semigroup generated by $\mathcal{H} \subset \mathbb{Z}^n$, then $H = \mathbb{Z}H \cap \mathbb{R}_+\mathcal{H}$.*

(b) *If C is a finitely generated rational cone in $\mathbb{R}^n$, then $H = \mathbb{Z}^n \cap C$ is a normal affine semigroup.*

Proof. (a) By (5.2), we only have to show that $H = \mathbb{Z}H \cap \mathbb{Q}_+\mathcal{H}$. Let $\mathbf{g} \in \mathbb{Z}H \cap \mathbb{Q}_+\mathcal{H}$, that is, $\mathbf{g} \in \mathbb{Z}H$ and $\mathbf{g} = a_1\mathbf{h}_1 + \cdots + a_q\mathbf{h}_q$ for some $a_1, \ldots, a_q \in \mathbb{Q}_+$. Then, by clearing the denominators in the previous equality, we may find a positive integer m such that $m\mathbf{g} \in H$. Since H is normal, it follows that $\mathbf{g} \in H$. This proves the inclusion $\mathbb{Z}H \cap \mathbb{Q}_+\mathcal{H} \subset H$. The other inclusion is trivial.

(b) We first show that H is a finitely generated semigroup. Since C is a rational finitely generated cone, there exist $\mathbf{q}_1, \ldots, \mathbf{q}_m \in \mathbb{Q}^n$ such that

$$C = \mathbb{R}_+\{\mathbf{q}_1, \ldots, \mathbf{q}_m\} = \{\sum_{i=1}^{m} a_i\mathbf{q}_i : a_i \in \mathbb{R}_+ \text{ for } 1 \leq i \leq m\}.$$

Multiplying $\mathbf{q}_1, \ldots, \mathbf{q}_m$ by a suitable factor, we may assume that $\mathbf{q}_1, \ldots, \mathbf{q}_m \in \mathbb{Z}^n$. Then, by (5.3), $\mathbb{Z}^n \cap C = \mathbb{Z}^n \cap \mathbb{Q}_+\{\mathbf{q}_1, \ldots, \mathbf{q}_m\}$. Let $\mathbf{c} \in \mathbb{Z}^n \cap C$. Then there exist $a_1, \ldots, a_m \in \mathbb{Q}_+$ such that $\mathbf{c} = \sum_{i=1}^m a_i \mathbf{q}_i$. Let $\mathbf{c} = \mathbf{c}' + \mathbf{c}''$, where $\mathbf{c}' = \sum_{i=1}^m a_i' \mathbf{q}_i$ with $a_i' \in \mathbb{N}$ for all i, and $\mathbf{c}'' = \sum_{i=1}^m a_i'' \mathbf{q}_i$ with $0 \leq a_i < 1$ for all i. Obviously, $\mathbf{c}' \in \mathbb{Z}^n \cap C$, hence $\mathbf{c}'' \in \mathbb{Z}^n$. Moreover, $\mathbf{c}''$ belongs to the bounded set

$$B = \{\sum_{i=1}^m a_i'' \mathbf{q}_i : 0 \leq a_i < 1 \text{ for } i = 1, \ldots, m\}.$$

$\mathbb{Z}^n \cap B$ is a finite set, since B is bounded. Then the set $(B \cap \mathbb{Z}^n) \cup \{\mathbf{q}_1, \ldots, \mathbf{q}_m\}$ is also finite and generates $\mathbb{Z}^n \cap C$, hence $H = \mathbb{Z}^n \cap C$ is finitely generated. The normality of H is immediate. □

We now proceed to show the equivalence between the normality of the affine semigroup and the associated semigroup ring.

Theorem 5.15. *Let H be an affine semigroup ring and K a field. Then $K[H]$ is normal if and only if H is normal.*

Proof. We have seen before that if $K[H]$ is a normal ring, then H is also normal. Now, let H be a normal affine semigroup generated by $\mathcal{H} = \{\mathbf{h}_1, \ldots, \mathbf{h}_q\} \subset \mathbb{Z}^n$, and let us write the cone $\mathbb{R}_+\mathcal{H}$ as intersection of half-spaces, let us say, $H_1^+, \ldots, H_m^+$. Then

$$H = \mathbb{Z}H \cap (H_1^+ \cap \cdots \cap H_m^+) = \bigcap_{i=1}^m (\mathbb{Z}H \cap H_i^+).$$

We set $L_i = \mathbb{Z}H \cap H_i^+$. Then $K[H] = \bigcap_{i=1}^m K[L_i]$, hence it is enough to show that $K[L_i]$ is normal for $1 \leq i \leq m$. But $L_i \cong \mathbb{Z}^{d-1} \oplus \mathbb{N}$ where $d = \operatorname{rank} \mathbb{Z}H$. Indeed, let

$$H_i = \{x \in \mathbb{R}^n : \langle \mathbf{c}_i, \mathbf{x} \rangle = 0\}$$

where $\mathbf{c}_i$ is an integer vector. Then the kernel of the map $\sigma : \mathbb{Z}H \to \mathbb{Z}$ given by $\sigma(\mathbf{x}) = \langle \mathbf{c}_i, \mathbf{x} \rangle$ is the group $\mathbb{Z}H \cap H_i$, hence it is isomorphic to $\mathbb{Z}^{d-1}$. Notice that this is also the kernel of the restriction of σ to L_i. On the other hand, the image of this restriction is a normal subsemigroup of $\mathbb{N}$, hence it is isomorphic to $\mathbb{N}$. Since the map $\mathbb{Z}H \to \operatorname{Im} \sigma$ splits, the restriction $L_i \to \operatorname{Im}(\sigma_{|L_i}) \cong \mathbb{N}$ splits as well and yields the desired presentation of L_i. Hence $K[L_i] \cong K[x_1, x_1^{-1}, \ldots, x_{d-1}, x_{d-1}^{-1}, x_d]$ which is the localization of a normal ring, therefore it is normal. □

From now on we make the assumption that the affine semigroup H is generated by a subset of $\mathbb{N}^n$. This condition is not very restrictive since any affine semigroup H with the property that $\mathbf{h} \in H$ and $-\mathbf{h} \in H$ implies $\mathbf{h} = 0$, can be embedded in $\mathbb{N}^n$. Our purpose is to investigate the connection between the normality of $K[H]$, equivalently of H, and the Gröbner basis of the toric ideal of $K[H]$.

Let H be generated by $\mathcal{H} = \{\mathbf{h}_1, \ldots, \mathbf{h}_q\} \subset \mathbb{N}^n$, let $R = K[t_1, \ldots, t_q]$ be the polynomial ring, and P_H the toric ideal of H, that is, $P_H = \operatorname{Ker} \varphi$ where $\varphi : R \to K[H]$ is defined by $t_i \mapsto \mathbf{x}^{\mathbf{h}_i}$ for $1 \leq i \leq q$.

Theorem 5.16 (Sturmfels). *Let $<$ be a monomial order on R such that $\mathtt{in}_<(P_H)$ is a squarefree monomial ideal. Then $K[H]$ is normal.*

Proof. Let Δ be the simplicial complex on the vertex set $[q]$ such that $I_\Delta = \mathtt{in}_<(P_H)$. Then, by Macaulay's Theorem (Theorem 2.6), we have the following decomposition of $K[H]$,

$$K[H] = \bigoplus_{F \in \Delta} \bigoplus_{\operatorname{supp}(\mathbf{a}) = F} K\mathbf{y}^{\mathbf{a}} = \sum_{F \in \mathcal{F}(\Delta)} \bigoplus_{\operatorname{supp}(\mathbf{a}) \subset F} K\mathbf{y}^{\mathbf{a}},$$

where $\mathbf{y}^{\mathbf{a}} = y_{i_1}^{a_{i_1}} \cdots y_{i_r}^{a_{i_r}}$ if $\operatorname{supp}(\mathbf{a}) = \{i_1, \ldots, i_r\}$ and $y_i = t_i + \mathtt{in}_<(P_H)$ for $1 \leq i \leq q$. This equality implies that $H = \bigcup_{F \in \mathcal{F}(\Delta)} H_F$ where H_F is the semigroup generated by $\mathbf{h}_{i_1}, \ldots, \mathbf{h}_{i_r}$ if $F = \{i_1, \ldots, i_r\}$. We note from the above decomposition of $K[H]$ that, for every $F \in \mathcal{F}(\Delta)$, $K[H_F]$ is a polynomial ring, hence it is normal. Let $\mathbf{g} \in \mathbb{Z}H$ such that $m\mathbf{g} \in H$ for some positive integer m. Then $m\mathbf{g} \in H_F$ for some $F \in \mathcal{F}(\Delta)$. Therefore, $\mathbf{g} + \mathbb{Z}H_F$ is a torsion element in $\mathbb{Z}H/\mathbb{Z}H_F$. By [**St95**, Chapter 8], the sets $\{\mathbf{h}_i : i \in F\}$ where $F \in \mathcal{F}(\Delta)$, form an unimodular regular triangulation of the set $\mathcal{H} = \{\mathbf{h}_1, \ldots, \mathbf{h}_q\} \subset \mathbb{N}^n$. This implies that $\mathbb{Z}H = \mathbb{Z}H_F$ for all $F \in \mathcal{F}(\Delta)$. Therefore, we have $g \in \mathbb{Z}H_F$. As we have seen before, H_F is a normal semigroup, hence $g \in H_F$, whence $g \in H$. □

The following theorem is due to Hochster [**Ho72**].

Theorem 5.17. *Let H be an affine normal semigroup and K a field. Then the semigroup ring $K[H]$ is Cohen-Macaulay.*

5.5. Edge rings associated with bipartite graphs

Well-known examples of toric ideals come from combinatorics. In this section we study semigroup rings associated with graphs which are called **edge rings**. We determine the Gröbner bases of the toric ideals of edge rings and show, as an application, that the edge rings associated with bipartite graphs are normal Cohen-Macaulay domains.

To begin with, we recall some basic terminology of graphs which will be also useful in the next chapter. A simple graph G is defined by its vertex set, say $[n] = \{1, \ldots, n\}$, and a set of edges, $E(G)$. Each edge is a subset $e \subset [n]$ with two distinct elements. A **walk** of length r of G is a subgraph

W of G with the edge set $E(W) = \{\{i_0, i_1\}, \{i_1, i_2\}, \ldots, \{i_{r-1}, i_r\}\}$ where $i_0, i_1, \ldots, i_r$ are vertices of G. A walk W with $i_0 = i_r$ is called **closed**. A **cycle** of length r is a closed walk $C = \{\{i_0, i_1\}, \{i_1, i_2\}, \ldots, \{i_{r-1}, i_r = i_0\}\}$ of length r where the vertices $i_0, i_1, \ldots, i_{r-1}$ are pairwise distinct. If r is even (odd), then the walk or cycle W is called an **even (odd) walk** or **cycle**. A **path** in G is a walk with pairwise distinct vertices. The graph G is **connected** if any two vertices are connected by a path. The maximal connected subgraphs of G are called the **connected components** of G.

Given a subset of vertices $\{i_1, \ldots, i_q\}$ of the graph G, the subgraph $G' = G \setminus \{i_1, \ldots, i_q\}$ on the vertex set $[n] \setminus \{i_1, \ldots, i_q\}$ which consists of all the edges of G not incident to any of the vertices $i_1, \ldots, i_q$, is called a **restriction** of G. One may also consider restrictions of G obtained by removing some of its edges. Namely, if $e_1, \ldots, e_q \in E(G)$, the restriction $H = G \setminus \{e_1, \ldots, e_q\}$ is defined on the same vertex set as G and it has the edge set $E(H) = E(G) \setminus \{e_1, \ldots, e_q\}$. A graph G is called **bipartite** if its vertex set can be partitioned into two nonempty and disjoint sets V_1 and V_2 such that any edge of G connects a vertex of V_1 to a vertex of V_2. A **complete graph** is a graph in which every pair of distinct vertices is connected by an edge.

Let G be a simple graph on the edge set $[n]$ and let $S = K[x_1, \ldots, x_n]$ the polynomial ring. With each edge $e = \{i, j\} \in E(G)$ we associate the quadratic squarefree monomial $\mathbf{x}_e = x_i x_j \in S$. Let $E(G) = \{e_1, \ldots, e_m\}$ be the edge set of G. The semigroup ring $K[G] = K[\mathbf{x}_{e_1}, \ldots, \mathbf{x}_{e_m}]$ is called the **edge ring** of G. Let $R = K[t_1, \ldots, t_m]$ be the polynomial ring in m indeterminates and P_G the toric ideal of $K[G]$, that is, the kernel of the surjective homomorphism $\varphi : R \to K[G]$ defined by $t_i \mapsto \mathbf{x}_{e_i}$ for $1 \le i \le m$.

With an even closed walk W of G with

$$E(W) = \{\{i_0, i_1\}, \{i_1, i_2\}, \ldots, \{i_{2r-1}, i_{2r} = i_0\}\},$$

we associate the binomial

$$f_W = \prod_{j=1}^{r} t_{e_{2j-1}} - \prod_{j=1}^{r} t_{e_{2j}} \in R$$

where $e_j = \{i_{j-1}, i_j\}$ for $1 \le j < 2r$ and $e_{2r} = \{i_{2r-1}, i_{2r} = i_0\}$. It is clear that $\varphi(f_W) = 0$, hence $f_W \in P_G$ for any even closed walk W of G.

An even closed walk W of G is called **primitive** if the associated binomial f_W is primitive. Obviously, an even cycle is a primitive walk. The converse is true for bipartite graphs.

Lemma 5.18. *Let G be a bipartite graph. Then every primitive walk of G is an even cycle.*

Proof. Let V_1, V_2 be the bipartition of the vertices of G and W a primitive walk with $E(W) = \{e_1, \ldots, e_{2r}\}$ where $e_j = \{i_{j-1}, i_j\}$ for $j = 1, \ldots, 2r$, and $i_0 = i_{2r}$. If W is not a cycle, since there are no edges with both vertices in the same set, V_1 or V_2, we may find $1 \leq \ell < p \leq r$ such that $i_{2\ell} = i_{2p}$ or $i_{2\ell-1} = i_{2p-1}$. In the first case we get

$$\mathbf{x}_{e_{2\ell+1}}\mathbf{x}_{e_{2\ell+3}} \cdots \mathbf{x}_{e_{2p-1}} = \mathbf{x}_{e_{2\ell+2}}\mathbf{x}_{e_{2\ell+4}} \cdots \mathbf{x}_{e_{2p}},$$

which implies that $t_{2\ell+1}t_{2\ell+3} \cdots t_{2p-1} - t_{2\ell+2}t_{2\ell+4} \cdots t_{2p} \in P_G$, a contradiction since f_W is a primitive binomial.

In the second case we get

$$\mathbf{x}_{e_{2\ell}}\mathbf{x}_{e_{2\ell+2}} \cdots \mathbf{x}_{e_{2p-2}} = \mathbf{x}_{e_{2\ell+1}}\mathbf{x}_{e_{2\ell+3}} \cdots \mathbf{x}_{e_{2p-1}},$$

which implies that $t_{2\ell+1}t_{2\ell+3} \cdots t_{2p-1} - t_{2\ell}t_{2\ell+2} \cdots t_{2p-2} \in P_G$, again a contradiction. □

Proposition 5.19. *Let G be a graph on the vertex set $[n]$. Then the set*

$$\mathcal{G} = \{f_W : W \text{ is a primitive walk in } G\}$$

is a Gröbner basis of P_G with respect to any monomial order.

Proof. By Proposition 5.8, it is enough to show that any primitive binomial $f \in P_G$ is of the form $f = f_W$ for some even closed walk W of G which will be primitive by definition. Let $f \in P_G$ be a primitive binomial. Without loss of generality, we may assume that $f = t_1t_3 \cdots t_{2r-1} - t_2t_4 \cdots t_{2r}$ for some integer $r \geq 2$. Then we have

$$\mathbf{x}_{e_1}\mathbf{x}_{e_3} \cdots \mathbf{x}_{e_{2r-1}} = \mathbf{x}_{e_2}\mathbf{x}_{e_4} \cdots \mathbf{x}_{e_{2r}}. \tag{5.4}$$

Let $e_1 = \{j_0, j_1\}$. Then $x_{j_1} | \mathbf{x}_{e_2} \cdots \mathbf{x}_{e_{2r}}$. Without loss of generality, we may assume that $e_2 = \{j_1, j_2\}$ for some $j_2 \neq j_1, j_0$. This implies, by equation (5.4), that $x_{j_2} | \mathbf{x}_{e_3} \cdots \mathbf{x}_{e_{2r-1}}$. We may assume that $e_3 = \{j_2, j_3\}$ for some $j_3 \neq j_2, j_1$. Thus, by (5.4), $x_{j_3} | \mathbf{x}_{e_4} \cdots \mathbf{x}_{e_{2r}}$. Let $e_4 = \{j_3, j_4\}$ for some $j_4 \neq j_3, j_2$. If $r = 2$, then equation (5.4) implies that $j_4 = j_0$ and $f = t_1t_3 - t_2t_4 = f_W$ where W is a primitive walk of length 4. If $r > 2$, then $j_4 \neq j_0$ since f is primitive. Equation (5.4) yields the equality $x_{j_0}\mathbf{x}_{e_5} \cdots \mathbf{x}_{e_{2r-1}} = x_{j_4}\mathbf{x}_{e_6} \cdots \mathbf{x}_{e_{2r}}$. Thus $x_{j_4} | \mathbf{x}_{e_5} \cdots \mathbf{x}_{e_{2r-1}}$. Let $e_5 = \{j_4, j_5\}$ for some $j_5 \neq j_4, j_3$. It follows that $x_{j_5} | \mathbf{x}_{e_6} \cdots \mathbf{x}_{e_{2r}}$. Let $e_6 = \{j_5, j_6\}$ for some $j_6 \neq j_5, j_4$. If $r = 3$, then, by using again equation (5.4), we get $j_6 = j_0$ and $f = t_1t_3t_5 - t_2t_4t_6 = f_W$ for a primitive walk W of length 6. If $r > 3$, then $j_6 \neq j_0$ and we continue the above procedure. Now it is clear that for any r we reach the conclusion that f is the binomial attached to a primitive walk. □

In particular, if the graph G is bipartite, we get that the set of primitive binomials associated with the even cycles of G forms a Gröbner basis of

P_G. Since the monomials of such a binomial are obviously squarefree, by applying Theorem 5.16 and Theorem 5.17, we get the following

Corollary 5.20. *Let G be a bipartite graph. Then its edge ring $K[G]$ is a normal Cohen-Macaulay domain.*

Problems

Problem 5.1. Find all the initial ideals for the toric ideal of the semigroup ring $K[x^3, x^2y, xy^2, y^3]$.

Problem 5.2. Give an example of a binomial ideal which is not a toric ideal.

Problem 5.3. Let c be a positive integer and $H = \{0\} \cup (c + \mathbb{N})$ where $c + \mathbb{N} = \{c + n : n \in \mathbb{N}\}$. Compute the toric ideal of $K[H]$.

Problem 5.4. Let $I \subset K[t_1, \dots, t_q]$ be an ideal with the property that $I : (t_i) = I$ for all $i = 1, \dots, q$. Show that $I : (t_1 \cdots t_q)^\infty = I$.

Problem 5.5. Let $L \subset \mathbb{Z}^q$ be a lattice and $I_L \subset K[t_1, \dots, t_q]$ its associated ideal. Show that any reduced Gröbner basis of I_L consists of binomials of the form $f = \mathbf{t}^\mathbf{u} - \mathbf{t}^\mathbf{v}$ where $\mathbf{u}, \mathbf{v} \in \mathbb{N}^q$ and $\mathbf{u} - \mathbf{v} \in L$.

Problem 5.6. Let $u_1, \dots, u_m$ be monomials in $K[x_1, x_1^{-1}, \dots, x_n, x_n^{-1}]$ with exponent vectors $\mathbf{v}_1, \dots, \mathbf{v}_m \in \mathbb{Z}^n$, and let $W \subset \mathbb{Z}^m$ be the lattice of vectors $\mathbf{w} = (w_1, \dots, w_m)$ with $\sum_{i=1}^m w_i \mathbf{v}_i = 0$. Then $\mathbf{t}^\mathbf{a} - \mathbf{t}^\mathbf{b}$ belongs to the toric ideal of the semigroup ring $K[u_1, \dots, u_m]$ if and only if $\mathbf{a} - \mathbf{b} \in W$.

Problem 5.7. Let $\mathcal{B}$ be a subset of $\mathbb{Z}^n$ and $I_\mathcal{B} \subset K[x_1, \dots, x_n]$ be the ideal generated by the binomials $f_\mathbf{b} = \mathbf{x}^{\mathbf{b}^+} - \mathbf{x}^{\mathbf{b}^-}, \mathbf{b} \in \mathcal{B}$. Let $g = \mathbf{x}^\mathbf{a} - \mathbf{x}^\mathbf{c} \in K[x_1, \dots, x_n]$ be a binomial. Show that $g \in I_\mathcal{B}$ if and only if there exists $\mathbf{d}_1, \dots, \mathbf{d}_m \in \mathbb{N}^n$ and $\mathbf{b}_1, \dots, \mathbf{b}_m \in \mathcal{B}$ such that $g = \sum_{i=1}^m \mathbf{x}^{\mathbf{d}_i}(\mathbf{x}^{\mathbf{b}_i^+} - \mathbf{x}^{\mathbf{b}_i^-})$.

Problem 5.8. Let $I \subset S = K[x_1, ..., x_n]$ be a binomial ideal, and let $f \in IS[x_{n+1}]$ be a primitive binomial. Then $f \in I$.

Problem 5.9. Let $I, J \subset S$ be graded ideals. Show that the sequence

$$0 \to \frac{S}{I \cap J} \xrightarrow{f} \frac{S}{I} \oplus \frac{S}{J} \xrightarrow{g} \frac{S}{I+J} \to 0$$

where $a + I \cap J \overset{f}{\mapsto} (a + I, -a + J)$ and $(a + I, b + J) \overset{g}{\mapsto} a + b + (I + J)$, is an exact sequence of graded S-modules and morphisms.

Problem 5.10. Let Δ be a pure simplicial complex. Show that Δ is shellable with respect to the order $F_1, \dots, F_m$ of its facets if and only if $\langle F_1, \dots, F_i \rangle$ is Cohen-Macaulay for all $i = 1, \dots, m$.

Problem 5.11. Show that the ring $K[x^3, x^2y, y^3]$ is not normal.

Problem 5.12. Let $H \subset \mathbb{N}$ be the semigroup generated by $\{3, 4, 5\}$. Show that $K[H]$ is a Cohen-Macaulay ring which is not normal.

Problem 5.13. Let $\mathcal{S} = d_1\mathbb{N} + \cdots + d_q\mathbb{N}$ be a semigroup where $d_1 < \cdots < d_q$ are positive integers. If $\gcd(d_1, \ldots, d_q) = 1$, then $\mathcal{S}$ is called a **numerical semigroup**. Show that $\mathcal{S}$ is a numerical semigroup if and only if there exists $c \in \mathcal{S}$ such that $c + \mathbb{N} \subset \mathcal{S}$.

Problem 5.14. Let $\mathcal{S} = d_1\mathbb{N} + \cdots + d_q\mathbb{N}$ be a semigroup where $d_1 < \cdots < d_q$ are positive integers. Show that $\mathcal{S}$ is normal if and only if $\mathcal{S}$ can be generated by a single element.

Problem 5.15. Let $d \geq 1$ be an integer and $\mathcal{M}_d$ the set of all the monomials of degree d of the polynomial ring $S = K[x_1, \ldots, x_n]$. Show that $K[\mathcal{M}_d]$ is a normal ring.

Problem 5.16. Let $R = K[t_1, \ldots, t_q]$ and let $f \in R$ be a primitive binomial which has a squarefree term. Show that the ring $R/(f)$ is a normal domain.

Problem 5.17. Let $S = K[x_1, \ldots, x_n]$ and let $K[H] \subset S$ be a semigroup ring. A polynomial $f \in S$ is called **integral** over $K[H]$ if there exists an integer $m > 1$ such that $f^m \in K[H]$. Show that the set of all the polynomials of S which are integral over $K[H]$ forms a semigroup ring. Determine this ring in the case $K[H] = K[x^a, y^b] \subset K[x, y]$ where a, b are some positive integers.

Problem 5.18. Let G be the triangle, that is, the graph on the vertex set $[3]$ with the edge set $E(G) = \{\{1, 2\}, \{2, 3\}, \{1, 3\}\}$. Show that the edge ring of G is normal.

Problem 5.19. Let $S_1, \ldots, S_r$ be polynomial rings over the field K on distinct sets of indeterminates and $K[H_i] \subset S_i$ for $i = 1, \ldots, r$, semigroup rings. Show that $K[\bigcup_{i=1}^{r} H_i]$ is a normal ring if and only if $K[H_i]$ is normal for all $1 \leq i \leq r$.

Chapter 6

Selected applications in commutative algebra and combinatorics

6.1. Koszul algebras

In this section we introduce the important class of Koszul algebras and show that a standard graded K-algebra $R = K[x_1, \dots, x_n]/I$ is Koszul if I has a quadratic Gröbner basis with respect to some monomial order.

A finitely generated graded module M over R has a (possibly infinite) minimal graded free R-resolution $\mathbb{F}$, where each F_i is of the form $F_i = \bigoplus_j R(-j)^{b_{ij}}$. Since the resolution is minimal, the numbers b_{ij} are uniquely determined by M; cf. the proof of Theorem 4.25. We set $\beta_{ij}(M) = b_{ij}$ for all i and j, and call these numbers the **graded Betti numbers** of M. Furthermore we set $\beta_i(M) = \sum_j \beta_{ij}(M)$.

Definition 6.1. *Let K be a field and R a standard graded K-algebra with graded maximal ideal $\mathfrak{m}$. The K-algebra R is called **Koszul**, if the residue class field $K = R/\mathfrak{m}$ has a linear R-resolution. In other words, if $\beta_{ij}(R/\mathfrak{m}) = 0$ for $j \neq i$.*

It follows that R is Koszul if and only if the entries of the matrices describing the maps in the minimal graded free R-resolution of $R/\mathfrak{m}$ are all elements of degree 1.

The formal power series $S_R(t) = \sum_{i \geq 0} \beta_i(R/\mathfrak{m})t^i$ is called the **Poincaré series** of R. It is known that, in general, this power series is not a rational function. But in the case of a Koszul algebra we have

Proposition 6.2. *Let R be a Koszul algebra. Then $S_R(-t)\operatorname{Hilb}_R(t) = 1$. In particular, $S_R(t)$ is a rational series of the form $(1+t)^d/P_R(t)$, where $d = \dim R$ and $P_R(t)$ is a polynomial with integer coefficients.*

Proof. If R is a Koszul algebra, then the graded free resolution of $R/\mathfrak{m}$ is of the form

$$\cdots \longrightarrow R(-2)^{\beta_2} \longrightarrow R(-1)^{\beta_1} \longrightarrow R \longrightarrow R/\mathfrak{m} \longrightarrow 0$$

with $\beta_i = \beta_i(R/\mathfrak{m})$ for all i. Since the Hilbert function is additive on short exact sequences we see that

$$\begin{aligned} 1 &= \operatorname{Hilb}_{R/\mathfrak{m}}(t) = \operatorname{Hilb}_R(t) - \beta_1 t \operatorname{Hilb}_R(t) + \beta_2 t^2 \operatorname{Hilb}_R(t) - \cdots \\ &= (1 - \beta_1 t + \beta_2 t^2 - \cdots)\operatorname{Hilb}_R(t) = S_R(-t)\operatorname{Hilb}_R(t), \end{aligned}$$

as desired. The remaining statements follow from Proposition 4.27. □

The simplest example of a Koszul algebra is the polynomial ring $S = K[x_1, \dots, x_n]$. Here the Koszul complex provides a (finite) linear resolution of the residue class ring; see [**HH10**, Appendix A.3]. The polynomial ring is the only graded K-algebra R for which the residue class field has a finite free R-resolution; see for example [**BH98**, Theorem 2.2.7].

By far, not every standard graded K-algebra is Koszul. Indeed one has

Proposition 6.3. *Let $R = S/I$ be a Koszul algebra with $I \subset (x_1, \dots, x_n)^2$. Then I is generated by polynomials of degree 2.*

Proof. Let $f_1, \dots, f_m$ be a minimal homogeneous system of generators of I, and let $U \subset F = \bigoplus_{i=1}^n Re_i$ be the kernel of the canonical epimorphism $\epsilon\colon F \to \mathfrak{m}$ with $\epsilon(e_i) = \bar{x}_i$ for $i = 1, \dots, n$. Here $\bar{f}$ denotes the residue class modulo I of a polynomial $f \in S$.

For $i = 1, \dots, m$, write $f_i = \sum_{j=1}^n f_{ij}x_j$ with homogeneous polynomials f_{ij}. We claim that U is generated by the homogeneous relations $u_i = \sum_{j=1}^n \bar{f}_{ij}e_j$ together with the $\bar{r}_{ij} = \bar{x}_i e_j - \bar{x}_j e_i$, $i < j$, and that none of the generators u_i can be omitted. The claim implies that the matrix describing the relations of $\mathfrak{m}$ has linear entries if and only if all f_i are of degree 2. This yields the desired conclusion.

In order to prove the claim we first notice that obviously all the elements u_i and $\bar{r}_{ij}$ belong to U. Now let $\sum_{j=1}^n \bar{g}_j e_j$ be an arbitrary element in U. Then $\sum_{j=1}^n \bar{g}_j \bar{x}_j = 0$, and so $\sum_{j=1}^n g_j x_j \in I$. Hence there exist $h_i \in S$ such that $\sum_{j=1}^n g_j x_j = \sum_{i=1}^m h_i f_i$. It follows that

$$\sum_{j=1}^n g_j x_j = \sum_{i=1}^m h_i\Big(\sum_{j=1}^n f_{ij}x_j\Big) = \sum_{j=1}^n\Big(\sum_{i=1}^m h_i f_{ij}\Big)x_j.$$

Consequently, $\sum_{j=1}^n(g_j - \sum_{i=1}^m h_i f_{ij})x_j = 0$. This implies that $\sum_{j=1}^n(g_j - \sum_{i=1}^m h_i f_{ij})e_j$ is an element of the kernel of the map $\bigoplus_{j=1}^n Se_j \to (x_1,\dots,x_n)$ with $e_j \mapsto x_j$ for $j = 1,\dots,n$. By Problem 6.1 this kernel is generated by the elements $r_{kl} = x_k e_l - x_l e_k$, $k < l$. Thus there exist polynomials p_{kl} such that

$$\sum_{j=1}^{n}(g_j - \sum_{i=1}^{m} h_i f_{ij})e_j = \sum_{k<l} p_{kl} r_{kl}.$$

It follows that $\sum_{j=1}^n \bar{g}_j e_j = \sum_{i=1}^m \bar{h}_i u_i + \sum_{k<l} \bar{p}_{kl}\bar{r}_{kl}$. This shows that the elements u_i and $\bar{r}_{kl}$ generate U.

Finally, suppose that one of the elements u_i, say u_1, can be omitted in the above generating set. Then there exist polynomials q_i and s_{kl} in S such that $u_1 = \sum_{i=2}^m \bar{q}_i u_i + \sum_{k<l} \bar{s}_{kl}\bar{r}_{kl}$. This equation implies that

$$\sum_{j=1}^{n} f_{1j}e_j - \sum_{i=2}^{m} q_i(\sum_{j=1}^{n} f_{ij}e_j) + \sum_{k<l} s_{kl}r_{kl} \in \bigoplus_{j=1}^{n} Ie_j.$$

Substituting the e_j by the x_j we obtain that $f_1 - \sum_{i=2}^m q_i f_i \in (x_1,\dots,x_n)I$, which by Nakayama's lemma is impossible since $f_1,\dots,f_m$ is a minimal system of generators of I. $\square$

Next we want to discover classes of Koszul algebras. To this end we introduce the following slightly stronger notion of Koszulness given in [**HHR00**].

Definition 6.4. *A standard graded K-algebra R is called* ***strongly Koszul*** *if its graded maximal ideal $\mathfrak{m}$ admits a minimal system of homogeneous generators $u_1,\dots,u_n$ such that for all subsequences $u_{i_1},\dots,u_{i_r}$ of $u_1,\dots,u_n$ and for all $j = 1,\dots,r-1$, the ideal quotient $(u_{i_1},\dots,u_{i_{j-1}}) : u_{i_j}$ is generated by a subset of elements of $\{u_1,\dots,u_n\}$.*

The naming "strongly Koszul" is justified by the following theorem.

We recall that a graded R-module M is **linear**, if it admits a system of generators $g_1,\dots,g_m$, all of the same degree, such that for $j = 1,\dots,m$ the ideal quotients

$$(Rg_1 + \cdots + Rg_{j-1}) : g_j = \{a \in R\colon\ ag_j \in Rg_1 + \cdots + Rg_{j-1}\}$$

are generated by subsets of $\{u_1,\dots,u_n\}$.

Theorem 6.5. *Let R be strongly Koszul with respect to the minimal homogeneous system $u_1,\dots,u_n$ of generators of the graded maximal ideal $\mathbf{m}$ of R. Then any ideal of the form $(u_{i_1},\dots,u_{i_r})$ has a linear resolution. In particular, R is Koszul.*

Proof. We first notice that all the ideals $(u_{i_1},\dots,u_{i_r})$ are linear modules. Hence the theorem will be proved if we have shown that a linear module M

has linear relations, and that the first syzygy module (relation module) of a linear module is again linear.

The first property is immediately clear. Indeed, if $a_1g_1+\cdots+a_mg_m$ is a homogeneous generating relation of M, and a_j is the last non-zero coefficient of this relation, then a_j is a generator of the ideal quotient $(Rg_1+\cdots+Rg_{j-1}):g_j$, and hence is of degree 1. Therefore, the relation is linear.

Let $\Omega^1(M)$ denote the first syzygy module of M. We prove by induction on the number of generators of M, that $\Omega^1(M)$ is a linear module. If M is cyclic, then $\Omega^1(M)$ is an ideal generated by a subset of $\{u_1,\dots,u_n\}$, and hence is linear. Now suppose that M is generated by the m elements $g_1,\dots,g_m$, $m>1$, for which M is linear. Then $N=Rg_1+\cdots+Rg_{m-1}$ is also linear, and by our induction hypothesis has a linear syzygy module $\Omega^1(N)$. Say, $\Omega^1(N)$ is linear with respect to its generators $h_1,\dots,h_k$. Now we build a suitable system of generators for $\Omega^1(M)$ using the exact sequence

$$0\longrightarrow\Omega^1(N)\longrightarrow\Omega^1(M)\longrightarrow\Omega^1(M/N)\longrightarrow 0.$$

The module M/N is cyclic with annihilator $(Rg_1+\cdots+Rg_{m-1}):g_m$, and so there exist $1\leq i_1<i_2<\ldots<i_l\leq n$ such that $\Omega^1(M/N)\simeq(u_{i_1},\dots,u_{i_l})$. Now we choose homogeneous elements $h_{k+1},\dots,h_{k+l}$ in $\Omega^1(M)$ mapping onto $u_{i_1},\dots,u_{i_l}$. Then $\Omega^1(M)$ is linear with respect to the generators $h_1,\dots,h_{k+l}$. □

From the preceding theorem we deduce the following remarkable result.

Corollary 6.6 (Fröberg). *Let R be the factor ring of a polynomial ring modulo an ideal which is generated by quadratic monomials. Then R is Koszul.*

Proof. We prove that R is strongly Koszul. Let $R=K[x_1,\dots,x_n]/I$, and denote the residue classes of the x_i by y_i. We have to show that for any sequence of elements $y_{i_1},\dots,y_{i_k}$, the ideal quotient $(y_{i_1},\dots,y_{i_{k-1}}):y_{i_k}$ is generated by a subsequence of $y_1,\dots,y_n$. We observe that

$$(6.1)\qquad (y_{i_1},\dots,y_{i_{k-1}}):y_{i_k}=[(I,x_{i_1},\dots,x_{i_{k-1}}):x_{i_k}]/I.$$

Let $I=(u_1,\dots,u_m)$ where the u_i are monomials of degree 2. Then

$$(I,x_{i_1},\dots,x_{i_{k-1}})=(u_{j_1},\dots,u_{j_r},x_{i_1},\dots,x_{i_{k-1}}),$$

where there u_{j_k} are those among the u_i which are not divisible by any of the variables $x_{i_1},\dots,x_{i_{k-1}}$. Now it follows from Proposition 1.14 that

$$(6.2)\qquad (I,x_{i_1},\dots,x_{i_{k-1}}):x_{i_k}$$
$$=\ (u_{j_1}/\gcd(u_{j_1},x_{i_k}),\dots,u_{j_r}/\gcd(u_{j_r},x_{i_k}),x_{i_1},\dots,x_{i_{k-1}}).$$

If x_{i_k} divides u_{j_s}, then $u_{j_s}/\gcd(u_{j_s},x_{i_k})$ is a variable because u_{j_s} is a monomial of degree 2, and if x_{i_k} does not divide u_{j_s}, then $u_{j_s}/\gcd(u_{j_s},x_{i_k})=u_{j_s}$.

Thus it follows from (6.1) and (6.2) that $(y_{i_1}, \ldots, y_{i_{k-1}}) : y_{i_k}$ is generated by a subset of $y_1, \ldots, y_n$, as desired. □

As usual, let $S = K[x_1, \ldots, x_n]$ be the polynomial ring over the field K in the indeterminates $x_1, \ldots, x_n$. We now come to the main result of this section.

Theorem 6.7. *Let $I \subset S$ be a graded ideal and assume that I has a quadratic Gröbner basis with respect to a monomial order on S. Then $R = S/I$ is Koszul.*

This theorem is an immediate consequence of Corollary 6.6 and the next result, as we shall see in a moment.

Theorem 6.8. *Let $I \subset J \subset S$ be graded ideals, and let $<$ be a monomial order on S. Then*

$$\beta_{ij}^{S/I}(S/J) \leq \beta_{ij}^{S/\operatorname{in}_<(I)}(S/\operatorname{in}_<(J)).$$

Proof of Theorem 6.7. Let J be the graded maximal ideal $(x_1, \ldots, x_n)$ of S. Then Theorem 6.8 implies that $\beta_{ij}^{R}(R/\mathfrak{m}) \leq \beta_{ij}^{R'}(R'/\mathfrak{m}')$ for all i and j, where $\mathfrak{m}$ denotes the graded maximal ideal of $R = S/I$ and $\mathfrak{m}'$ the graded maximal ideal of $R' = S/\operatorname{in}_<(I)$. By Fröberg's result (Corollary 6.6) it follows that $\beta_{ij}^{R'}(R'/\mathfrak{m}') = 0$ for $j \neq i$. Thus $\beta_{ij}^{R}(R/\mathfrak{m}) = 0$ for $j \neq i$, which implies that R is Koszul. □

The proof of Theorem 6.8 needs some preparation. Given an integer vector $\mathbf{w} = (w_1, \ldots, w_n) \in \mathbb{N}^n$ we define a new grading on $S = K[x_1, \ldots, x_n]$ by setting $\deg_{\mathbf{w}}(x_i) = w_i$ for $i = 1, \ldots, n$. Then $\deg_{\mathbf{w}} \mathbf{x}^{\mathbf{a}} = \langle \mathbf{a}, \mathbf{w} \rangle$ where $\langle\ ,\ \rangle$ denotes the standard scalar product in $\mathbb{R}^n$. Let $f = \sum_i c_i \mathbf{x}^{\mathbf{a}_i} \in S$. We set $\deg_{\mathbf{w}} f = \max\{\langle \mathbf{a}_i, \mathbf{w} \rangle : c_i \neq 0\}$, and call the polynomial

$$f^h = \sum_i c_i \mathbf{x}^{\mathbf{a}_i} t^{\deg_{\mathbf{w}} f - \deg_{\mathbf{w}} \mathbf{x}^{\mathbf{a}_i}} \in S[t]$$

the **homogenization** of f with respect to $\mathbf{w}$. This definition coincides with the definition given in Chapter 3 in the case that $\mathbf{w} = (1, 1, \ldots, 1)$. Finally, similarly as in Chapter 3, for any ideal $I \subset S$ we set

$$I^h = (f^h\colon\ f \in I),$$

and call I^h the **homogenization** of I with respect to the weight $\mathbf{w}$.

In the sequel we shall use the following fundamental facts which can be found for example in [**HH10**, Chapter 3]:

(a) Let $I \subset S$ be a standard graded ideal, and fix a weight $\mathbf{w} \in \mathbb{N}^n$. We assign to each x_i the bidegree $\deg x_i = (w_i, 1)$ and to t the bidegree $(1, 0)$. Then

(i) I^h is bihomogeneous. In other words, I^h is generated by bihomogeneous polynomials of $S[t]$;

(ii) $S[t]/I^h$ is a free $K[t]$-module with respect to the natural K-algebra homomorphism.

(iii) $(S[t]/I^h)/(t-1)(S[t]/I^h) \cong S/I$.

(b) Let $I_1, \dots, I_r \subset S$ be standard graded ideals and $<$ a monomial order on S. Then there exists a weight vector $\mathbf{w}$ such that for the homogenizations of the I_j with respect to $\mathbf{w}$ we have that $\overline{(I_j)^h} = \mathtt{in}_<(I_j)$ for $j = 1, \dots, r$, where for an ideal $J \subset S[t]$ we denote by $\overline{J}$ its reduction modulo t.

Proof of Theorem 6.8. Let I^h be the homogenization of I and J^h be the homogenization of J with respect to a suitable weight such that $\overline{I^h} = \mathtt{in}_<(I)$ and $\overline{J^h} = \mathtt{in}_<(J)$. Let $\mathbb{F}$ by the bigraded minimal free $S[t]/I^h$-resolution of $S[t]/J^h$. Then each F_i is of the form $\bigoplus_{k,j} S[t]/I^h(-k,-j)^{b_{i,kj}}$. We set $F_{ij} = \bigoplus_k S[t]/I^h(-k,-j)^{b_{i,kj}}$, so that $F_i = \bigoplus_j F_{ij}$. Since $S[t]/I^h$ and $S[t]/J^h$ are free $K[t]$-modules, it follows that t and $t-1$ are nonzerodivisors on $\mathbb{F}$ and $S[t]/J^h$. Therefore $\mathbb{F}/t\mathbb{F}$ is a free $S/\mathtt{in}_<(I)$-resolution of $S/\mathtt{in}_<(J)$, and $\mathbb{F}/(t-1)\mathbb{F}$ is a free S/I-resolution of S/J. Both resolutions are graded, because t and $t-1$ respect the second component of the bidegree. The first resolution is minimal, because t belongs to the graded maximal ideal of $S[t]$. The second resolution may not be minimal. Thus we see that

$$\begin{aligned} \beta_{ij}^{S/I}(S/J) &\le \operatorname{rank}_{S/I} F_{ij}/(t-1)F_{ij} = \operatorname{rank}_{S[t]/I^h} F_{ij} \\ &= \operatorname{rank}_{S/\mathtt{in}_<(I)} F_{ij}/tF_{ij} = \beta_{ij}^{S/\mathtt{in}_<(I)}(S/\mathtt{in}_<(J)), \end{aligned}$$

as desired. □

Let $R = S/I$ where I is a graded ideal. Summarizing we have the following implications:

I has a quadratic Gröbner basis $\Rightarrow$ R is Koszul $\Rightarrow$ I is generated by quadrics.

All these implications are strict as the following examples due to Hibi and Ohsugi [**HO99**, Example 2.2 and Example 2.1] show:

(a) The ideal $I = (x_2x_8 - x_4x_7, x_1x_6 - x_3x_5, x_1x_3 - x_2x_4) \subset K[x_1, \dots, x_8]$ has no quadratic Gröbner basis. However, the ring $K[x_1, \dots, x_8]/I$ is Koszul.

(b) The ideal $(x_4x_6 - x_5x_9, x_3x_{10} - x_4x_8, x_2x_9 - x_3x_7, x_1x_{10} - x_5x_7, x_1x_8 - x_2x_6)$ is the toric ideal attached to a graph whose toric ring is not Koszul.

Classes of Koszul algebras will be considered in the following sections.

We close this section with another important consequence of Theorem 6.8. Here we refer to the definitions given in Subsection 4.4.6.

Corollary 6.9. *Let $R = S/I$ be a standard graded K-algebra, and $<$ a monomial order on S with the property that $S/\mathtt{in}_<(I)$ is Cohen–Macaulay (Gorenstein). Then R is Cohen–Macaulay (Gorenstein).*

6.2. Sortable sets of monomials

We have seen in Section 5.4 that if the toric ideal of an affine semigroup H has a squarefree initial ideal, then the semigroup ring $K[H]$ is normal and Cohen–Macaulay. In addition, if $K[H]$ is standard graded and its toric ideal has a quadratic Gröbner basis, then $K[H]$ is Koszul (see Corollary 6.6). Therefore, it is of interest to find classes of toric ideals which have squarefree initial ideals or quadratic Gröbner bases. For this purpose, we need to study sortable sets of monomials, a concept which is due to Sturmfels [**St95**].

Let d be a positive integer, S_d the K-vector space generated by the monomials of degree d in $S = K[x_1, \dots, x_n]$, and take two monomials $u, v \in S_d$. We write $uv = x_{i_1}x_{i_2}\cdots x_{i_{2d}}$ with $1 \le i_1 \le i_2 \le \cdots \le i_{2d} \le n$ and define

$$u' = x_{i_1}x_{i_3}\cdots x_{i_{2d-1}}, v' = x_{i_2}x_{i_4}\cdots x_{i_{2d}}.$$

The pair (u', v') is called the **sorting** of (u, v). In this way we obtain a map

$$\mathrm{sort} : S_d \times S_d \to S_d \times S_d, (u, v) \mapsto (u', v').$$

This map is called the **sorting operator**. For example, if $u = x_1x_3^2$ and $v = x_2^2x_3$, then $\mathrm{sort}(u, v) = (x_1x_2x_3, x_2x_3^2)$. A pair (u, v) is **sorted** if $\mathrm{sort}(u, v) = (u, v)$. Notice that $\mathrm{sort}(u, v) = \mathrm{sort}(v, u)$, and that if (u, v) is sorted, then $u \ge_{\mathrm{lex}} v$.

Definition 6.10. A subset $B \subset S_d$ of monomials is called **sortable** if $\mathrm{sort}(B \times B) \subset B \times B$.

A significant example of a sortable set is the following set of monomials. Let $\mathbf{a} = (a_1, \dots, a_n) \in \mathbb{N}^n$ be a vector and consider the set

$$S_{n,d}^{\mathbf{a}} = \{u = x_1^{c_1}\cdots x_n^{c_n} : \deg(u) = d, c_i \le a_i \text{ for } 1 \le i \le n\}.$$

The algebra $K[S_{n,d}^{\mathbf{a}}]$ is called of **Veronese type**. If $\mathbf{a} = (d, \dots, d)$, then $K[S_{n,d}^{\mathbf{a}}]$ is the dth Veronese subring of S, and if $\mathbf{a} = (1, \dots, 1)$, then we obtain the dth squarefree Veronese subalgebra of S.

Proposition 6.11. *The set $S_{n,d}^{\mathbf{a}}$ is sortable.*

Proof. Let $u, v \in S_{n,d}^{\mathbf{a}}$ and $(u', v') = \mathrm{sort}(u, v)$. We have to check that for every $1 \le i \le n$, the degree of x_i in u' and v' is at most a_i. Let $\deg_{x_i}(u) = b_i$ and $\deg_{x_i}(v) = c_i$. We have $\max\{b_i, c_i\} \le a_i$. By the definition of the sorting operator, if $b_i + c_i$ is even, then $\deg_{x_i}(u') = \deg_{x_i}(v') = (b_i + c_i)/2$, and if $b_i + c_i$ is odd, then $\{\deg_{x_i}(u'), \deg_{x_i}(v')\} = \{(b_i + c_i + 1)/2, (b_i + c_i - 1)/2\}$. In both cases we have $\deg_{x_i}(u'), \deg_{x_i}(v') \le a_i$. $\square$

An r-tuple of monomials $(u_1,\dots,u_r)\in S_d^r$ is called **sorted** if for any $1\le i<j\le n$, the pair (u_i,u_j) is sorted. In other words, if we write the monomials $u_1,\dots,u_r$ as $u_1=x_{i_1}\cdots x_{i_d}$, $u_2=x_{j_1}\cdots x_{j_d},\dots,u_r=x_{\ell_1}\cdots x_{\ell_d}$, then $(u_1,\dots,u_r)$ is sorted if and only if

$$(6.3)\quad i_1\le j_1\le\cdots\le \ell_1\le i_2\le j_2\le\cdots\le\ell_2\le\cdots\le i_d\le j_d\le\cdots\le\ell_d.$$

Indeed, from the above sequence of inequalities it clearly follows that every pair (u_i,u_j) with $i<j$ is sorted. Conversely, let $(u_1,\dots,u_r)$ be sorted. In particular, the pair (u_1,u_2) is sorted, hence $i_1\le j_1\le i_2\le j_2\le\cdots\le i_d\le j_d$. Let $u_3=x_{k_1}\cdots x_{k_d}$ where $1\le k_1\le\cdots\le k_d\le n$. Since (u_2,u_3) is sorted, we have $j_1\le k_1\le j_2\le k_2\le\cdots\le j_d\le k_d$. On the other hand, (u_1,u_3) is sorted as well, thus we get $i_1\le j_1\le k_1\le i_2\le j_2\le k_2\le\cdots\le i_d\le j_d\le k_d$. Therefore, step by step, we get all the inequalities from (6.3).

Let $(u_1,\dots,u_r)$ be an r-tuple such that for some $i<j$ the pair (u_i,u_j) is unsorted, and let $(u_i',u_j')=\operatorname{sort}(u_i,u_j)$. Then we call the assignment $(u_1,\dots,u_r)\mapsto(u_1,\dots,u_i',\dots,u_j',\dots,u_r)$ a **single sorting step**.

Theorem 6.12. *Let $(u_1,\dots,u_r)$ be an r-tuple of monomials in S_d.*
(a) *There is a unique sorted r-tuple $(u_1',\dots,u_r')$ such that $u_1\cdots u_r=u_1'\cdots u_r'$.*
(b) *The unique sorted r-tuple can be obtained from the original one in a finite number of single sorting steps.*

Proof. (a) If $u_1u_2\cdots u_r=x_{i_1}x_{i_2}\cdots x_{i_{rd}}$ with $1\le i_1\le i_2\le\cdots\le i_{rd}\le n$, then $u_j'=\prod_{k=0}^{d-1}x_{i_{rk+j}}$ for $1\le j\le r$.

(b) We write $u_1=x_{i_1}\cdots x_{i_d}$, $1\le i_1\le\cdots\le i_d\le n$, $u_2=x_{j_1}\cdots x_{j_d}$, $1\le j_1\le\cdots\le j_d\le n,\dots,$ and $u_r=x_{\ell_1}\cdots x_{\ell_d}$, $1\le\ell_1\le\cdots\le\ell_d\le n$. Then

$$u_1u_2\cdots u_r=(x_{i_1}x_{j_1}\cdots x_{\ell_1})(x_{i_2}x_{j_2}\cdots x_{\ell_2})\cdots(x_{i_d}x_{j_d}\cdots x_{\ell_d}).$$

We relabel the sequence $i_1,j_1,\dots,\ell_1,i_2,j_2,\dots,\ell_2,\dots,i_d,j_d,\dots,\ell_d$ as

$$\mathbf{k}=k_1,k_2,\dots,k_r,k_{r+1},k_{r+2},\dots,k_{2r},\dots,k_{(d-1)r+1},k_{(d-1)r+2},\dots,k_{dr}.$$

We define a function s from the set of all r-tuples of monomials of degree d to $\mathbb{Z}$ which is given by $s(\mathbf{u})=\sum_{i<j}(k_j-k_i)$. Obviously, this function is bounded above. The proof is completed once we have shown that $s(\mathbf{u})<s(\mathbf{u}')$ where $\mathbf{u}'$ is obtained from $\mathbf{u}$ by a single sorting step. Suppose that the pair (u_i,u_j) is unsorted. The pair (u_i,u_j) defines a subsequence of $\mathbf{k}$, let us say $k_{i_1},\dots,k_{i_{2d}}$, and let $\mathcal{S}=\{i_1,\dots,i_{2d}\}$. We split the sum defining $s(\mathbf{u})$ as follows:

$$s(\mathbf{u})=\sum_{\substack{i<j\\ i,j\in\mathcal{S}}}(k_j-k_i)+\sum_{\substack{i<j\\ i,j\notin\mathcal{S}}}(k_j-k_i)+\sum_{j\notin\mathcal{S}}\left[\sum_{\substack{i<j\\ i\in\mathcal{S}}}(k_j-k_i)+\sum_{\substack{i>j\\ i\in\mathcal{S}}}(k_i-k_j)\right].$$

We apply the same decomposition for $s(\mathbf{u}')$ with respect to the pair (u_i', u_j'). Note that the set of indices defining the corresponding subsequence for (u_i', u_j') in $\mathbf{k}'$ is again $\mathcal{S} = \{i_1, \ldots, i_{2d}\}$. Comparing the sums in the decompositions for $s(\mathbf{u})$ and $s(\mathbf{u}',)$ we see that first sum becomes strictly larger, the second sum does not change, and for each $j \notin \mathcal{S}$, the sums in the square brackets do not become smaller. The first part of the above claim follows in the following way. Let $\mathbf{l} = \ell_1, \ldots, \ell_r$ be a sequence of positive integers and $\mathbf{l}' = \ell_1', \ldots, \ell_r'$ a permutation of $\mathbf{l}$ such that $\ell_1' \leq \cdots \leq \ell_r'$. Let $f(\mathbf{l}) = \sum_{i<j}(\ell_j - \ell_i)$, $g(\mathbf{l}) = \sum_{i<j}|\ell_j - \ell_i|$, and $g(\mathbf{l}') = \sum_{i<j}|\ell_j' - \ell_i'| = \sum_{i<j}(\ell_j' - \ell_i')$. We notice that $g(\mathbf{l}) = g(\mathbf{l}')$. Indeed, it is clear that $g(\mathbf{l})$ and $g(\mathbf{l}')$ have the same number of summands. Let now $|\ell_j - \ell_i|$ be a summand of $g(\mathbf{l})$. Then there exist uniquely determined integers p, q such that $\ell_j = \ell_p'$ and $\ell_i = \ell_q'$. If $p < q$, then $|\ell_j - \ell_i| = \ell_q' - \ell_p'$ and if $p > q$, then $|\ell_j - \ell_i| = \ell_p' - \ell_q'$. Thus there exists a bijection between the summands of $g(\mathbf{l})$ and the summands of $g(\mathbf{l}')$. We obviously have the inequality $f(\mathbf{l}) \leq g(\mathbf{l})$, thus $f(\mathbf{l}) \leq g(\mathbf{l}')$. Moreover, the equality in the last inequality holds if and only if $f(\mathbf{l}) = g(\mathbf{l})$, that is, $\ell_i \leq \ell_j$ for all $i < j$.

Finally, we have to compare the last sums in the decompositions of $s(\mathbf{u})$ and $s(\mathbf{u}')$. For a fixed $j \notin \mathcal{S}$, let $\mathcal{S}^< = \{i \in \mathcal{S} : i < j\}$ and $\mathcal{S}^> = \{i \in \mathcal{S} : i > j\}$. Then, for $j \notin \mathcal{S}$, the inequality

$$\sum_{i\in\mathcal{S}^<}(k_j - k_i) + \sum_{i\in\mathcal{S}^>}(k_i - k_j) \geq \sum_{i\in\mathcal{S}^<}(k_j - k_i') + \sum_{i\in\mathcal{S}^>}(k_i' - k_j)$$

is equivalent to

$$\sum_{i\in\mathcal{S}^<} k_i - \sum_{i\in\mathcal{S}^>} k_i \geq \sum_{i\in\mathcal{S}^<} k_i' - \sum_{i\in\mathcal{S}^>} k_i'.$$

Since $\sum_{i\in\mathcal{S}} k_i = \sum_{i\in\mathcal{S}} k_i'$, the last inequality is equivalent to $\sum_{i\in\mathcal{S}^<} k_i \geq \sum_{i\in\mathcal{S}^<} k_i'$, which obviously holds by the construction of the sorting of a pair of monomials. □

Let $B \subset S_d$ be a sortable set of monomials and $K[B]$ the semigroup ring generated over K by B. Let $R = K[\{t_u : u \in B\}]$ be the polynomial ring with the order of indeterminates given by $t_u > t_v$ if $u >_{\text{lex}} v$ and $\varphi : R \to K[B]$ the K-algebra homomorphism defined by $t_u \mapsto u$ for all $u \in B$. Let P_B be the kernel of φ. We are interested in studying the Gröbner basis of P_B with respect to a suitable monomial order. We need the following

Definition 6.13. Let $\mathcal{F} = \{f_1, \ldots, f_s\} \subset R$ be a finite family of marked binomials in R. In other words, $f_i = \underline{m_i} - m_i'$ for some monomials $m_i, m_i' \in R$ for all $i = 1, \ldots, s$, where the monomial m_i in f_i is marked by underline. $\mathcal{F}$ is called **marked coherently** if there exists a monomial order $<$ on R such that $\mathtt{in}_<(f_i) = m_i$ for $1 \leq i \leq s$.

The next theorem, due to Sturmfels [**St95**], gives a necessary and sufficient condition for a finite set $\mathcal{F} = \{f_1, \dots, f_s\}$ of marked binomials to be marked coherently. In order to recall this theorem we need one more definition. Let $g \in R$ be a nonzero polynomial, and let w be one of the monomials of the support of g. If w is divisible by the marked monomial m_i of $f_i = \underline{m_i} - m_i' \in \mathcal{F}$, we replace the factor m_i in w by m_i'. We let h be the new polynomial obtained in this way and say that g **reduces to** h **modulo** $\mathcal{F}$.

Theorem 6.14. *Let $\mathcal{F} \subset R$ be a set of marked binomials. Then $\mathcal{F}$ is marked coherently if and only if every sequence of reductions modulo $\mathcal{F}$ terminates in a finite number of steps.*

With these tools at hand we have the following result.

Theorem 6.15. *Let B be a sortable subset of monomials of S_d and*

$$\mathcal{F} = \{\underline{t_u t_v} - t_{u'} t_{v'} : u, v \in B, (u, v) \textit{ unsorted, } (u', v') = \mathrm{sort}(u, v)\}.$$

Then there exists a monomial order $<$ on R which is called the ***sorting order*** *such that for every $g = \underline{t_u t_v} - t_{u'} t_{v'} \in \mathcal{F}$, $\mathtt{in}_<(g) = t_u t_v$.*

Proof. By Theorem 6.12, any r-tuple of monomials $(u_1, \dots, u_r)$ can be sorted by a finite number of sorting steps. This statement is equivalent to saying that our set $\mathcal{F}$ is marked coherently by Theorem 6.14. □

Finally, as the main result of this subsection we obtain

Theorem 6.16. *Let $K[B]$ be the K-algebra generated by a sortable set of monomials $B \subset S_d$ and $P_B \subset R$ its toric ideal. Then*

$$\mathcal{G} = \{t_u t_v - t_{u'} t_{v'} : u, v \in B, (u, v) \textit{ unsorted, } (u', v') = \mathrm{sort}(u, v)\}$$

is the reduced Gröbner basis of P_B with respect to the sorting order.

Proof. By Theorem 6.15, $(\mathtt{in}_<(\mathcal{G}))$ is generated by the monomials $t_u t_v$ where (u, v) is unsorted. Therefore, the K-basis of $R/(\mathtt{in}_<(\mathcal{G}))$ is the set X of the monomials $t_{u_1} \cdots t_{u_r}$ of R with $(u_1, \dots, u_r)$ sorted. By Proposition 4.29 we only need to show that the set X forms a K-basis of R/P_B. By Theorem 6.12, every polynomial of R reduces to a finite K-linear combination of monomials in X. Therefore, the set X generates R/P_B over K. Next, since any linear combination of distinct monomials in X maps to a linear combination of distinct monomials in $K[B]$ it follows that X is K-linearly independent. □

There are examples which show that in general the sorting order is neither lexicographic nor reverse lexicographic [**St95**, Proposition 14.4.].

It follows from Proposition 6.11 and Theorem 6.16 that the toric ideals of Veronese type algebras have quadratic Gröbner bases with respect to the sorting order.

If we choose the reverse lexicographic order induced by the natural order of the indeterminates, then the Gröbner basis of the Veronese type algebras is no longer quadratic. For example, choose $S^{\mathbf{a}}_{3,3} \subset K[x_1, x_2, x_3]$ where $\mathbf{a} = (2, 2, 2)$. Then

$$S^{\mathbf{a}}_{3,3} = \{x_1^2x_2, x_1^2x_3, x_1x_2^2, x_1x_2x_3, x_1x_3^2, x_2^2x_3, x_2x_3^2\}.$$

Let $P \subset K[t_1, \ldots, t_7]$ be the presentation ideal of the semigroup ring associated with $S^{\mathbf{a}}_{3,3}$. Its reduced Gröbner basis with respect to the reverse lexicographic order is $\mathcal{G}_{\text{revlex}} = \{t_1t_5^2 - t_2^2t_7, t_1t_6^2 - t_3^2t_7, t_2t_3 - t_1t_4, t_2t_4 - t_1t_5, t_3t_4 - t_1t_6, t_4^2 - t_1t_7, t_3t_5 - t_1t_7, t_4t_5 - t_2t_7, t_2t_6 - t_1t_7, t_4t_6 - t_3t_7, t_5t_6 - t_4t_7\}$.

However, we can show that the degree of the binomials in the reduced Gröbner basis for Veronese type algebras with respect to the reverse lexicographic order is at most three.

Theorem 6.17. *Let $\mathcal{G}$ be the reduced Gröbner basis of the toric ideal of $K[S^{\mathbf{a}}_{n,d}]$ with respect to the reverse lexicographic order induced by $t_u > t_v$ if $u >_{\text{lex}} v$. Then, every binomial $g \in \mathcal{G}$ has degree at most* 3.

Proof. Let $g = t_{u_1} \cdots t_{u_q} - t_{v_1} \cdots t_{v_q} \in \mathcal{G}$ with $\mathtt{in}_<(g) = t_{u_1} \cdots t_{u_q}$. Since g is primitive, hence irreducible, it follows that $t_{u_q} > t_{v_q}$, that is, $u_q >_{\text{lex}} v_q$. Therefore, there exists an index ℓ such that $\deg_{x_1}(u_q) = \deg_{x_1}(v_q), \ldots, \deg_{x_{\ell-1}}(u_q) = \deg_{x_{\ell-1}}(v_q)$, and $\deg_{x_\ell}(u_q) > \deg_{x_\ell}(v_q)$. But u_q and v_q have the same degree, thus there exists $h > \ell$ such that $\deg_{x_h}(u_q) < \deg_{x_h}(v_q)$. On the other hand, $u_1 \cdots u_q = v_1 \cdots v_q$, thus, by comparing the degrees of each indeterminate in both products, we may find $1 \le a, b \le q-1$ such that

$$\deg_{x_\ell}(u_a) < \deg_{x_\ell}(v_a) \text{ and } \deg_{x_h}(u_b) > \deg_{x_h}(v_b).$$

If $u_a = u_b$, then we set $u'_a = x_\ell u_a/x_h$ and $u'_q = x_h u_q/x_\ell$. It follows that $h = t_{u_a}t_{u_q} - t_{u'_a}t_{u'_q} \in P_H$ and $t_{u_q} > t_{u'_q}$, whence $\mathtt{in}_<(h) = t_{u_a}t_{u_q}$. Since $\mathcal{G}$ is a reduced Gröbner basis we have $\mathtt{in}_<(g) = \mathtt{in}_<(h)$, which implies that $\deg(g) = 2$.

Finally, if $u_a \neq u_b$, there exists j such that $\deg_{x_j}(u_a) > \deg_{x_j}(u_b.)$ In this case, we consider the monomials

$$u'_a = x_\ell u_a/x_j, u'_b = x_j u_b/x_h, \text{ and } u'_q = x_h u_q/x_\ell.$$

It follows that $u_a u_b u_q = u'_a u'_b u'_q$, thus $h = t_{u_a}t_{u_b}t_{u_q} - t_{u'_a}t_{u'_b}t_{u'_q} \in P_H$, and $u_q >_{\text{lex}} u'_q$, hence $\mathtt{in}_<(h) = t_{u_a}t_{u_b}t_{u_q}$. Since $\mathcal{G}$ is reduced, it follows that $\mathtt{in}_<(g)$ is a monomial of degree at most 3. $\square$

6.3. Generalized Hibi rings

In 1985 Hibi [**Hi87**] introduced a class of algebras which nowadays are called **Hibi rings**. They are semigroup rings attached to finite posets, and may be viewed as natural generalizations of polynomial rings. Indeed, a polynomial ring in n variables over a field K is just the Hibi ring of the poset $[n] = \{1, 2, \ldots, n\}$.

Hibi rings appear naturally in various combinatorial and algebraic contexts, for example, in invariant theory.

Let $P = \{p_1, \ldots, p_n\}$ be a finite poset. A **poset ideal** I of P is a subset of P which satisfies the following condition. For every $p \in I$, if $q \in P$ and $q \leq p$, then $q \in I$. Let $\mathcal{I}(P)$ be the set of the poset ideals of P. It is easily seen that $\mathcal{I}(P)$ is a sublattice of the power set of P, hence it is a distributive lattice. By Birkhoff's theorem any finite distributive lattice arises in this way. Let K be a field. Then the Hibi ring over K attached to P is the toric ring $K[\mathcal{I}(P)]$ generated by the set of monomials $\{x_I t\colon\ I \in \mathcal{I}(P)\}$ where $x_I = \prod_{p_i \in I} x_i$. Let $T = K[\{t_I\colon\ t_I \in \mathcal{I}(P)\}]$ be the polynomial ring in the variables t_I over K, and $\varphi\colon T \to K[\mathcal{I}(P)]$ the K-algebra homomorphism with $t_I \mapsto x_I t$. One fundamental result concerning Hibi rings is that the toric ideal $L_P = \operatorname{Ker} \varphi$ has a reduced Gröbner basis consisting of the so-called **Hibi relations**:

$$t_I t_J - t_{I \cap J} t_{I \cup J} \quad \text{with} \quad I \not\subseteq J \quad \text{and} \quad J \not\subseteq I.$$

Hibi showed [**Hi87**] that any Hibi ring is a normal Cohen–Macaulay domain, and that it is Gorenstein if and only if the attached poset P is pure, that is, all maximal chains of P have the same cardinality.

More generally, for any lattice $\mathcal{L}$, not necessarily distributive, one may consider the K algebra $K[\mathcal{L}]$ with generators y_α, $\alpha \in \mathcal{L}$, and relations $y_\alpha y_\beta = y_{\alpha \wedge \beta} y_{\alpha \vee \beta}$ where $\wedge$ and $\vee$ denote meet and join in $\mathcal{L}$. Hibi showed that $K[\mathcal{L}]$ is a domain if and only if $\mathcal{L}$ is distributive, in other words, if $\mathcal{L}$ is an ideal lattice of a poset.

Hibi ideals were first introduced in [**HH05**]. To each finite poset $P = \{p_1, \ldots, p_n\}$, one may attach the **Hibi ideal** H_P as the monomial ideal in the polynomial ring with $2n$ indeterminates $K[x_1, \ldots, x_n, y_1, \ldots, y_n]$ generated by the monomials $x_I y_{P \setminus I}$ with $I \in \mathcal{I}(P)$. Note that the toric ring generated over K by these monomials is isomorphic to the Hibi ring attached to P.

We now present the theory of generalized Hibi rings as introduced in [**EHM10**].

Let $\mathcal{I}(P)$ be the set of poset ideals of P and r a positive integer. An **r-multichain** of $\mathcal{I}(P)$ is a chain of poset ideals of length r,

$$\mathcal{I} : I_1 \subseteq I_2 \subseteq \cdots \subseteq I_r = P.$$

We define a partial order on the set $\mathcal{I}_r(P)$ of all r-multichains of $\mathcal{I}(P)$ by setting $\mathcal{I} \leq \mathcal{I}'$ if $I_k \subseteq I'_k$ for $k = 1, \ldots, r$. Observe that the partially ordered set $\mathcal{I}_r(P)$ is a distributive lattice, if we define the meet of $\mathcal{I}$: $I_1 \subseteq \cdots \subseteq I_r$ and $\mathcal{I}'$: $I'_1 \subseteq \cdots \subseteq I'_r$ as $\mathcal{I} \cap \mathcal{I}'$ where $(\mathcal{I} \cap \mathcal{I}')_k = I_k \cap I'_k$ for $k = 1, \ldots, r$, and the join as $\mathcal{I} \cup \mathcal{I}'$ where $(\mathcal{I} \cup \mathcal{I}')_k = I_k \cup I'_k$ for $k = 1, \ldots, r$.

With each r-multichain of $\mathcal{I}_r(P)$ we associate a monomial $u_\mathcal{I}$ in the polynomial ring $S = K[\{x_{ij} : 1 \leq i \leq r, 1 \leq j \leq n\}]$ in rn indeterminates which is defined as

$$u_\mathcal{I} = x_{1J_1} x_{2J_2} \cdots x_{rJ_r},$$

where $x_{kJ_k} = \prod_{p_\ell \in J_k} x_{k\ell}$ and $J_k = I_k \setminus I_{k-1}$ for $k = 1, \ldots, r$.

Lemma 6.18. *Let $\mathcal{I}$ and $\mathcal{I}'$ be two r-multichains of $\mathcal{I}(P)$. Then*

$$u_\mathcal{I} u_{\mathcal{I}'} = u_{\mathcal{I} \cup \mathcal{I}'} u_{\mathcal{I} \cap \mathcal{I}'}.$$

Proof. Indeed, the equality holds if and only if

$$\frac{x_{tI_t}}{x_{tI_{t-1}}} \cdot \frac{x_{tI'_t}}{x_{tI'_{t-1}}} = \frac{x_{tI_t \cap I'_t}}{x_{tI_{t-1} \cap I'_{t-1}}} \cdot \frac{x_{tI_t \cup I'_t}}{x_{tI_{t-1} \cup I'_{t-1}}}$$

for $t = 1, \ldots, r$.

In order to see that this identity holds, just observe that

$$x_{tI_t \cap I'_t} = \gcd\{x_{tI_t}, x_{tI'_t}\}, \quad x_{tI_{t-1} \cap I'_{t-1}} = \gcd\{x_{tI_{t-1}}, x_{tI'_{t-1}}\},$$

and

$$x_{tI_t \cup I'_t} = \frac{x_{tI_t} \cdot x_{tI'_t}}{\gcd\{x_{tI_t}, x_{tI'_t}\}} \quad x_{tI_{t-1} \cup I'_{t-1}} = \frac{x_{tI_{t-1}} \cdot x_{tI'_{t-1}}}{\gcd\{x_{tI_{t-1}}, x_{tI'_{t-1}}\}}.$$

□

Theorem 6.19. *The set of monomials $\{u_\mathcal{I} : \mathcal{I} \in \mathcal{I}_r(P)\} \subset S_{rn}$ is sorted with respect to $x_{11} > x_{21} > \cdots > x_{r1} > x_{12} > \cdots > x_{r2} > \cdots > x_{1n} > \cdots > x_{rn}$.*

Proof. Let $\mathcal{I}, \mathcal{I}' \in \mathcal{I}_r(P)$ be two r-multichains. We show that, with respect to the given order of the indeterminates, $\operatorname{sort}(u_\mathcal{I}, u_{\mathcal{I}'}) = (u_{\mathcal{I} \cup \mathcal{I}'}, u_{\mathcal{I} \cap \mathcal{I}'})$. By Lemma 6.18, we have $u_\mathcal{I} u_{\mathcal{I}'} = u_{\mathcal{I} \cup \mathcal{I}'} u_{\mathcal{I} \cap \mathcal{I}'}$. Next, we notice that for every $1 \leq j \leq n$, there exist uniquely determined $1 \leq k, \ell \leq r$ such that $p_j \in I_k \setminus I_{k-1}$ and $p_j \in I'_\ell \setminus I'_{\ell-1}$. Therefore, $x_{\ell j}$ and x_{kj} are the unique indeterminates with second index j which divide the product $u_\mathcal{I} u_{\mathcal{I}'}$. If $k = \ell$, then obviously x_{kj} divides $u_{\mathcal{I} \cup \mathcal{I}'}$ and $u_{\mathcal{I} \cap \mathcal{I}'}$. Let $k < \ell$. By the definition of the sorting and the chosen order of indeterminates, the conclusion follows once we show that $x_{kj} | u_{\mathcal{I} \cup \mathcal{I}'}$ and $x_{\ell j} | u_{\mathcal{I} \cap \mathcal{I}'}$. Since $k < \ell$ and $p_j \in I_k$, we obtain $p_j \in I_\ell$ as well. Therefore, we get $p_j \in (I_\ell \cap I'_\ell) \setminus (I_{\ell-1} \cap I'_{\ell-1})$ since $p_j \notin I'_{\ell-1}$. We thus have $x_{\ell j} | u_{\mathcal{I} \cap \mathcal{I}'}$. On the other hand, as $p_j \notin I'_{\ell-1}$ and $k < \ell$, it follows that $p_j \notin I'_{k-1}$, thus $p_j \in (I_k \cup I'_k) \setminus (I_{k-1} \cup I'_{k-1})$. Therefore, $x_{kj} | u_{\mathcal{I} \cup \mathcal{I}'}$ and the proof is completed. □

Let $R_r(P)$ be the K-subalgebra of S generated by the set $\{u_{\mathcal{I}} : \mathcal{I} \in \mathcal{I}_r(P)\}$. The ring $R_r(P)$ is called a **generalized Hibi ring**. This naming is justified by the fact that for $r = 2$ one obtains the classical Hibi ring. Let T be the polynomial ring over K in the set of indeterminates $\{t_{\mathcal{I}} : \mathcal{I} \in \mathcal{I}_r(P)\}$. Furthermore, let $\varphi\colon T \to R_r(P)$ be the surjective K-algebra homomorphism with $\varphi(t_{\mathcal{I}}) = u_{\mathcal{I}}$ for all $\mathcal{I} \in \mathcal{I}_r(P)$.

By applying Theorem 6.16 and Theorem 6.19 we get the following

Theorem 6.20. *The set*

$$\mathcal{G} = \{t_{\mathcal{I}} t_{\mathcal{I}'} - t_{\mathcal{I}\cup\mathcal{I}'} t_{\mathcal{I}\cap\mathcal{I}'} \in T : \mathcal{I}, \mathcal{I}' \in \mathcal{I}_r(P) \text{ incomparable}\},$$

is a reduced Gröbner basis of the ideal $L_r = \operatorname{Ker}\varphi$ with respect to the sorting order on T induced by $x_{11} > x_{21} > \cdots > x_{r1} > x_{12} > \cdots > x_{r2} > \cdots > x_{1n} > \cdots > x_{rn}$.

Corollary 6.21. *For any poset P and all integers $r \geq 1$, the toric ring $R_r(P)$ is a normal Cohen–Macaulay domain.*

It is interesting that the set of binomials which gives the reduced Gröbner basis of L_r with respect to the sorting order coincides with the reduced Gröbner basis with respect to the reverse lexicographic order induced by a total order of indeterminates with the property that $\mathcal{I} < \mathcal{I}'$ implies that $t_{\mathcal{I}} > t_{\mathcal{I}'}$.

Theorem 6.22. *The set*

$$\mathcal{G} = \{t_{\mathcal{I}} t_{\mathcal{I}'} - t_{\mathcal{I}\cup\mathcal{I}'} t_{\mathcal{I}\cap\mathcal{I}'} \in T : \mathcal{I}, \mathcal{I}' \in \mathcal{I}_r(P) \text{ incomparable}\},$$

is a reduced Gröbner basis of the ideal $L_r = \operatorname{Ker}\varphi$ with respect to the reverse lexicographic order induced by the given order of the variables $t_{\mathcal{I}}$.

Proof. Let $\prod_{s=1}^q t_{\mathcal{I}_s} - \prod_{s=1}^q t_{\mathcal{I}'_s}$ be a primitive binomial in L_r with initial monomial $\prod_{s=1}^q t_{\mathcal{I}_s}$. We are going to show that there are two indices k and ℓ such that $\mathcal{I}_k$ and $\mathcal{I}_\ell$ are incomparable r-multichains of ideals, and that $t_{\mathcal{I}_k} t_{\mathcal{I}_\ell}$ is the leading monomial of $t_{\mathcal{I}_k} t_{\mathcal{I}_\ell} - t_{\mathcal{I}_k\cup\mathcal{I}_\ell} t_{\mathcal{I}_k\cap\mathcal{I}_\ell}$. This will then show that $\mathcal{G}$ is Gröbner basis of L_r. It is obvious that $\mathcal{G}$ is actually reduced.

Suppose to the contrary that $\mathcal{I}_1 \leq \mathcal{I}_2 \leq \ldots \leq \mathcal{I}_q$. We will show that $\mathcal{I}'_s < \mathcal{I}_q$ for all s. Indeed, since $\prod_{s=1}^q t_{\mathcal{I}_s} - \prod_{s=1}^q t_{\mathcal{I}'_s} \in L_r$ we see that $\prod_{s=1}^q u_{\mathcal{I}_s} = \prod_{s=1}^q u_{\mathcal{I}'_s}$. It follows that

$$\prod_{s=1}^{q}\Big(\prod_{k=1}^{\ell} x_{kI_{sk}\setminus I_{sk-1}}\Big) = \prod_{s=1}^{q}\Big(\prod_{k=1}^{\ell} x_{kI'_{sk}\setminus I'_{sk-1}}\Big) \quad \text{for all } \ell = 1, \ldots, r.$$

Here $\mathcal{I}_s$ is the r-multichain of ideals $I_{s1} \subseteq I_{s2} \subseteq \cdots \subseteq I_{sr} = P$, and $\mathcal{I}'_s$ the r-multichain of ideals $I'_{s1} \subseteq I'_{s2} \subseteq \cdots \subseteq I'_{sr} = P$.

Now for all j and k we apply the substitution $x_{kj} \mapsto x_j$, and obtain

$$\prod_{s=1}^{q} x_{I_{s\ell}} = \prod_{s=1}^{q} x_{I'_{s\ell}}, \quad \ell = 1, \ldots, r,$$

where $x_J = \prod_{j \in J} x_j$ for $J \subset [n]$.

Since $\mathcal{I}_1 \leq \mathcal{I}_2 \leq \cdots \leq \mathcal{I}_q$, it follows that $\mathrm{supp}(\prod_{s=1}^{q} x_{I_{s\ell}}) = I_{q\ell}$. Thus the equation $\prod_{s=1}^{q} x_{I_{s\ell}} = \prod_{s=1}^{q} x_{I'_{s\ell}}$ implies that $x_{I'_{s\ell}} | x_{I_{q\ell}}$ for all ℓ and all s. It follows that $\mathcal{I}'_s \leq \mathcal{I}_q$. We cannot have equality, since $\prod_{s=1}^{q} t_{\mathcal{I}_s} - \prod_{s=1}^{q} t_{\mathcal{I}'_s}$ is a primitive binomial. This contradicts the fact that $\prod_{s=1}^{q} t_{\mathcal{I}_s}$ is the initial monomial of $\prod_{s=1}^{q} t_{\mathcal{I}_s} - \prod_{s=1}^{q} t_{\mathcal{I}'_s}$.

Finally, $t_{\mathcal{I}_k} t_{\mathcal{I}_\ell}$ is the leading monomial of $t_{\mathcal{I}_k} t_{\mathcal{I}_\ell} - t_{\mathcal{I}_k \cup \mathcal{I}_\ell} t_{\mathcal{I}_k \cap \mathcal{I}_\ell}$ thanks to the monomial order on T. □

6.4. Gröbner bases for Rees rings

6.4.1. The ℓ-exchange property. This subsection is devoted to the study of the Gröbner bases of presentation ideals of Rees rings defined by monomial ideals.

Let $I \subset S = K[x_1, \ldots, x_n]$ be a graded ideal. Recall from Example 3.10 that the Rees ring of $I = (f_1, \ldots f_m)$, denoted by $\mathcal{R}(I)$, is the graded subring of $S[t]$ given by

$$\mathcal{R}(I) = \bigoplus_{j \geq 0} I^j t^j = S[f_1 t, \ldots, f_m t].$$

The Rees ring $\mathcal{R}(I)$ has the presentation

$$\varphi : R = S[y_1, \ldots, y_m] \longrightarrow \mathcal{R}(I),$$

defined by

$$x_i \mapsto x_i \quad \text{for} \quad 1 \leq i \leq n \quad \text{and} \quad y_j \mapsto f_j t \quad \text{for} \quad 1 \leq j \leq m.$$

The ideal $J = \mathrm{Ker}(\varphi) \subset S[y_1, \ldots, y_m]$ is called **the presentation ideal** of $\mathcal{R}(I)$.

In the following we concentrate on the case that $I = (u_1, \ldots, u_m)$ is a monomial ideal generated in one degree. In this case R is a polynomial ring which admits a natural bigraded K-algebra structure which is given by setting $\deg(x_i) = (1, 0)$ for $i = 1, \ldots, n$ and $\deg(y_j) = (0, 1)$ for $j = 1, \ldots, m$.

On the other hand, let $T = K[y_1, \ldots, y_m]$ and L be the toric ideal of $K[u_1, \ldots, u_m]$ which is the kernel of the surjective homomorphism

$$\psi : T \to K[u_1, \ldots, u_m]$$

defined by $\psi(y_i) = u_i$ for all i.

Notice that $K[u_1, \dots, u_m]$ is isomorphic to the fiber $\mathcal{R}(I)/\mathfrak{m}\mathcal{R}(I)$ of $\mathcal{R}(I)$ since all generators of I have the same degree. Here $\mathfrak{m} = (x_1, \dots, x_n)$ is the graded maximal ideal of S.

Let $<$ be a monomial order on T. A monomial $\mathbf{y}^{\mathbf{a}}$ in T is called a **standard monomial** of L with respect to $<$, if it does not belong to the initial ideal of L.

The following definition appears first in [**HHV05**].

Definition 6.23. *The monomial ideal I satisfies the **ℓ-exchange property** with respect to the monomial order $<$ on T, if the following condition is satisfied: let $\mathbf{y}^a$ and $\mathbf{y}^b$ be any two standard monomials of L with respect to $<$ of the same degree with $u = \psi(\mathbf{y}^{\mathbf{a}})$ and $v = \psi(\mathbf{y}^{\mathbf{b}})$ satisfying*

(i) $\deg_{x_t} u = \deg_{x_t} v$ *for* $t = 1, \dots, q-1$ *with* $q \le n-1$,

(ii) $\deg_{x_q} u < \deg_{x_q} v$.

Then there exists an integer k, and an integer $q < j \le n$ such that $x_q u_k / x_j \in I$.

The usefulness of this concept becomes apparent in the next theorem which is taken from [**HHV05**].

Let $<'$ be an arbitrary monomial order on T and $<_{\text{lex}}$ the lexicographic order on S with respect to $x_1 > \cdots > x_n$. A new monomial order $<'_{\text{lex}}$ is defined on R as follows: for two monomials $\mathbf{x}^{\mathbf{a}}\mathbf{y}^{\mathbf{b}}$ and $\mathbf{x}^{\mathbf{c}}\mathbf{y}^{\mathbf{d}}$ in R, we set $\mathbf{x}^{\mathbf{a}}\mathbf{y}^{\mathbf{b}} <'_{\text{lex}} \mathbf{x}^{\mathbf{c}}\mathbf{y}^{\mathbf{d}}$ if and only if (i) $\mathbf{x}^{\mathbf{a}} <_{\text{lex}} \mathbf{x}^{\mathbf{c}}$ or (ii) $\mathbf{x}^{\mathbf{a}} = \mathbf{x}^{\mathbf{c}}$ and $\mathbf{y}^{\mathbf{b}} <' \mathbf{y}^{\mathbf{d}}$.

Recall that $<'_{\text{lex}}$ is just the product order of $<'$ and $<_{\text{lex}}$ as introduced in Chapter 2.

Theorem 6.24. *Let $I = (u_1, \dots, u_m)$ be a monomial ideal generated in one degree, satisfying the ℓ-exchange property. Then the reduced Gröbner basis of the toric ideal J with respect to $<'_{\text{lex}}$ consists of all binomials belonging to the reduced Gröbner basis of L with respect to $<'$ together with the binomials*

$$x_i y_k - x_j y_l,$$

where $x_i > x_j$ with $x_i u_k = x_j u_l$ and x_j is the smallest variable for which $x_i u_k / x_j$ belongs to I.

Proof. Let $\mathcal{G}$ denote the finite set which consists of all binomials belonging to the reduced Gröbner basis of L with respect to $<'$ and the binomials $x_i y_k - x_j y_l$, as described in the theorem.

Our goal is to show that $\mathcal{G}$ is a Gröbner basis of J with respect to $<'_{\text{lex}}$. Let $f \in R$ be an irreducible binomial belonging to J. If $\mathtt{in}_{<'_{lex}}(f) \in T$, then $f \in J \cap T = L$ and $\mathtt{in}_{<'_{lex}}(f)$ is divided by the initial monomial of a binomial belonging to the Gröbner basis of L with respect to $<'$. This is due to the fact that $<'_{\text{lex}}$ is an elimination order for $x_1, \ldots, x_n$. Next, suppose that $\mathtt{in}_{<'_{lex}}(f) \notin T$ and write $f = x_i\mathbf{x}^{\mathbf{a}}\mathbf{y}^{\mathbf{b}} - x_j\mathbf{x}^{\mathbf{a}'}\mathbf{y}^{\mathbf{b}'}$, where x_i is the biggest variable appearing in f and where $i < j$. We may assume that $\mathbf{y}^{\mathbf{b}}$ is a standard monomial of L with respect to $<'$. We will show that $x_i\mathbf{y}^{\mathbf{b}}$ is divided by the initial monomial of a binomial of type $x_iy_k - x_jy_l$, as described in the theorem.

Replacing $\mathbf{y}^{\mathbf{b}'}$ with its standard monomial of L with respect to $<'$, we may assume that both $\mathbf{y}^{\mathbf{b}}$ and $\mathbf{y}^{\mathbf{b}'}$ are standard monomials of L with respect to $<'$.

Since none of the variables x_k with $k < i$ appears in f, it follows that, for each $1 \leq k < i$, the power of the variable x_k appearing in the monomial $u = \psi(\mathbf{y}^{\mathbf{b}})$ is equal to the power of the variable x_k appearing in $v = \psi(\mathbf{y}^{\mathbf{b}'})$. In other words, $\deg_{x_k} u = \deg_{x_k} v$ for $k = 1, 2, \ldots, i-1$. On the other hand, since the variable x_i does not appear in $\mathbf{x}^{\mathbf{a}'}$, we also have $\deg_{x_i} u < \deg_{x_i} v$.

The ℓ-exchange property of I with respect to $<'$ yields the existence of integers s and $i < \ell \leq n$ such that $x_i(u_s/x_\ell) \in I$. Say, $x_i(u_s/x_\ell) = u_r$. Then $x_iy_s - x_\ell y_r \in J$ and its initial monomial divides $x_i\mathbf{y}^{\mathbf{b}}$. This shows that $\mathcal{G}$ is indeed a Gröbner basis of J. It is easy to see that it is a reduced Gröbner basis. ∎

6.4.2. The Rees ring of generalized Hibi ideals. In this subsection we use the ℓ-exchange property to compute the Gröbner basis of the presentation ideal of the Rees ring of generalized Hibi ideals.

The following proposition provides classes of ideals which satisfy the ℓ-exchange property. To describe the result we need to introduce the following concept due to M. Kokubo and T. Hibi [**KH06**].

Definition 6.25. *A squarefree monomial ideal I which is generated in one degree is called* ***weakly polymatroidal*** *with respect to the order $x_1 > x_2 > \cdots > x_n$ of the variables, if for any two generators $u = x_{i_1} \cdots x_{i_d}$ with $i_1 < i_2 < \cdots < i_d$, and $v = x_{j_1} \cdots x_{j_d}$ with $j_1 < j_2 < \cdots < j_d$ of I such that $i_1 = j_1, \ldots, i_{t-1} = j_{t-1}$ and $i_t < j_t$, there exists $\ell \geq t$ such that $x_{i_t}(v/x_{j_\ell}) \in I$.*

Now we have

Proposition 6.26. *Let $I \subset K[x_1, \ldots, x_n]$ be a weakly polymatroidal ideal which is sortable. Then I satisfies the ℓ-exchange property with respect to the sorting order.*

Proof. Let $I = (u_1, \ldots, u_m)$, L the toric ideal of $K[u_1, \ldots, u_m]$ and $\mathbf{y}^a$ and $\mathbf{y}^b$ be two standard monomials of L with respect to the sorting order satisfying (i) and (ii) of Definition 6.23. Suppose that $\psi(\mathbf{y}^a) = u_{i_1} \cdots u_{i_d}$ and $\psi(\mathbf{y}^b) = u_{j_1} \cdots u_{j_d}$, and that all pairs $(u_{i_l}, u_{i_{l'}})$ and $(u_{j_l}, u_{j_{l'}})$ are sorted. It follows from (i) that $\deg_{x_t}(u_{i_l}) = \deg_{x_t}(u_{j_l})$ for $l = 1, \ldots, d$ and for $t = 1, \ldots, q-1$, and (ii) implies that there exists $1 \leq l \leq d$ with $\deg_{x_q}(u_{i_l}) < \deg_{x_q}(u_{j_l})$. Since $\deg_{x_t}(u_{i_l}) = \deg_{x_t}(u_{j_l})$ for $t = 1, \ldots, q-1$ and $\deg_{x_q}(u_{i_l}) < \deg_{x_q}(u_{j_l})$, and since I is weakly polymatroidal there exists $j > q$ with $x_q u_{i_l}/x_j \in I$, as desired. □

Let P be a finite poset and $H_r(P) \subset K[\{x_{ij} : i = 1, \ldots, r, j = 1, \ldots, n\}]$ the rth generalized Hibi ideal as introduced in Section 6.3.

Proposition 6.27. *The ideal $H_r(P)$ is weakly polymatroidal with respect to the order*

$$x_{11} > x_{12} > \cdots > x_{1n} > x_{21} > \cdots > x_{2n} > \cdots > x_{r1} > \cdots > x_{rn}$$

of the variables.

Proof. Let $\mathcal{I} \colon I_1 \subseteq I_2 \subseteq \cdots \subseteq I_r = P$ and $\mathcal{J} \colon J_1 \subseteq J_2 \subseteq \cdots \subseteq J_r = P$ be two different chains of poset ideals and $u_{\mathcal{I}}$ and $u_{\mathcal{J}}$ the corresponding generators of $H_r(P)$. Let k be the first index such that $I_k \neq J_k$, and let $I_k \backslash I_{k-1} = \{p_{i_1}, \ldots, p_{i_a}\}$ with $i_1 < i_2 < \ldots < i_a$ and $J_k \backslash J_{k-1} = \{p_{j_1}, \ldots, p_{j_b}\}$ with $j_1 < j_2 < \ldots < j_b$. We may assume that there exists t such that $i_1 = j_1, \ldots, i_{t-t} = j_{t-1}$ and $i_t < j_t$. By our assumption on k, there exists $\ell > k$ such that $p_{i_t} \in J_\ell \setminus J_{\ell-1}$. It follows that $x_{ki_t}(u_{\mathcal{J}}/x_{\ell i_t}) \in H_r(P)$, because $x_{ki_t}(u_{\mathcal{J}}/x_{\ell i_t}) = u_{\mathcal{L}}$ where $\mathcal{L}$ is the chain

$$J_1 \subseteq J_2 \subseteq \cdots \subseteq J_k \cup \{p_{i_t}\} \subseteq \cdots \subseteq J_{\ell-1} \cup \{p_{i_t}\} \subseteq J_\ell \subseteq \cdots \subseteq J_r = P$$

of poset ideals of P. This shows that $H_r(P)$ is indeed weakly polymatroidal. □

Finally, as a consequence of Theorem 6.19, Proposition 6.27, Proposition 6.26 and Theorem 6.24 we obtain

Theorem 6.28. *Let $<'$ be the sorting order given by the sorting of the monomial generators $u_1, \ldots, u_m$ of $H_r(P)$. Then the reduced Gröbner basis of the presentation ideal of $\mathcal{R}(H_r(P))$ with respect to $<'_{\text{lex}}$ consists of all binomials belonging to the reduced Gröbner basis of the toric ideal L of the fiber $\mathcal{R}(H_r(P))/\mathfrak{m}\mathcal{R}(H_r(P))$ together with the binomials*

$$x_{ir}y_k - x_{js}y_l,$$

where $x_{ir} > x_{js}$ with $x_{ir}u_k = x_{js}u_l$ and x_{js} is the smallest variable for which $x_{ir}u_k/x_{js}$ belongs to $H_r(P)$.

Corollary 6.29. *The Rees ring $\mathcal{R}(H_r(P))$ is Koszul and a normal Cohen–Macaulay domain.*

Proof. Since the fiber of $\mathcal{R}(H_r(P))$ is generated by a sortable set of monomials, Theorem 6.16 implies that the reduced Gröbner basis of its toric ideal L is generated by binomials of degree 2 with squarefree leading terms. By Theorem 6.28, the reduced Gröbner basis of the presentation ideal J of $\mathcal{R}(H_r(P))$ is composed of the reduced Gröbner basis of L and of binomials of the type $x_{ir}y_k - x_{js}y_l$. Therefore the reduced Gröbner basis J consists of binomials of degree 2 with squarefree leading terms. Hence the assertions follow from Theorem 6.7 and Theorem 5.16. □

6.5. Determinantal ideals

This section is based on the paper [**HT96**] and the article [**BC03**]. There the interested reader can find more details related to determinantal ideals as well as extensions of the results presented here. A classical reference for determinantal rings are the lecture notes [**BV88**] by Bruns and Vetter. Here we approach the theory of determinantal ideals via Gröbner bases. To compute the Gröbner bases of these ideals we follow the elegant method of [**St90**].

6.5.1. Determinantal ideals and their initial ideals. In this section we compute a Gröbner basis for the ideal of t-minors of an $m \times n$-matrix of indeterminates. The proof presented here is due to Sturmfels [**St90**].

Let K be a field and $X = (x_{ij})_{\substack{i=1,\dots,m \\ j=1,\dots,n}}$ a matrix of indeterminates. We denote by $K[X]$ the polynomial ring over K with indeterminates x_{ij}.

Let $1 \leq t \leq \min\{m, n\}$ be an integer. We fix two sequences of integers,

$$1 \leq a_1 < a_2 < \cdots < a_t \leq m \quad \text{and} \quad 1 \leq b_1 < b_2 < \cdots < b_t \leq n,$$

and consider the determinant

$$[a_1 \cdots a_t | b_1 \cdots b_t] := \det \begin{pmatrix} x_{a_1b_1} & x_{a_1b_2} & \cdots & x_{a_1b_t} \\ x_{a_2b_1} & x_{a_2b_2} & \cdots & x_{a_2b_t} \\ \vdots & \vdots & & \vdots \\ x_{a_tb_1} & x_{a_tb_2} & \cdots & x_{a_tb_t} \end{pmatrix}.$$

Any such determinant is called a ***t*-minor** of X. The ideal generated by all t-minors of X will be denoted $I_t(X)$.

There is a natural partial order on the set $M(X)$ of all the minors of X defined as follows:

$$[a_1 \cdots a_t | b_1 \cdots b_t] \leq [c_1 \cdots c_s | d_1 \cdots d_s],$$

if and only if $t \geq s$, $a_i \leq c_i$ and $b_i \leq d_i$ for $i = 1, \dots, s$.

A product of minors $\delta = \delta_1\delta_2 \cdots \delta_r$ with $\delta_1 \leq \delta_2 \leq \cdots \leq \delta_r$ is called a **standard monomial**.

One calls an array A of positive integers (a_{ij}) with $1 \leq i \leq r$ and $1 < j < s_i$ such that $s_1 \geq s_2 \geq \cdots \geq s_r$ with strictly increasing rows and nondecreasing columns a **standard (Young) tableau**. The **shape** of the tableau A is the sequence $(s_1, \ldots, s_r)$ and $s_1 + \cdots + s_r$ is called its **degree**. A **bitableau** is an ordered pair $T = (A, B)$ of tableaux of same shape, and we set $\deg T = \deg A (= \deg B)$.

A standard monomial can be naturally identified with a bitableau. For example, if $\delta = \delta_1\delta_2\delta_3$ with $\delta_1 = [124|123]$, $\delta_2 = [13|23]$ and $\delta_3 = [24|25]$, then the bitableau associated with the standard monomial δ is given in Figure 1.

1	2	4
1	3	
2	4	

1	2	3
2	3	
2	5	

Figure 1. A bitableau

The following remarkable fact is a consequence of the straightening law of Doubilet-Rota-Stein; see for example [**BC03**] for the proof.

Theorem 6.30. *Let t be an integer with $1 \leq t \leq \min\{m, n\}$. Then the standard monomials $\delta = \delta_1\delta_2 \cdots \delta_r$ with $\deg \delta_1 \geq t$ form a K-basis of $I_t(X)$.*

For the computation of a Gröbner basis of $I_t(X)$ we apply the Knuth-Robinson-Schensted correspondence. For this purpose we first describe Schensted's deletion algorithm applied to a standard tableau A of shape $(s_1, \ldots, s_r)$: the input of DELETE is a positive integer $k \leq r$ with $s_k > s_{k+1}$, the output is an element a of A and a standard tableau $B = (b_{ij})$ of shape $(s_l, \ldots, s_k - 1, \ldots, s_r)$.

Set $a_k = a_{ks_k}$, and for $i = k-1, \ldots, 1$, let a_i be the largest element $\leq a_{i+1}$ in the ith row of A. Then $a = a_1$, and the standard tableau B is obtained from A by replacing a_i by a_{i+1} in the ith row for $i = 1, \ldots, k-1$ and by deleting the element a_{ks_k} in the kth row of A.

In order to demonstrate this algorithm we apply DELETE for $k = 3$ to the tableau in Figure 2.

The Knuth-Robinson-Schensted correspondence assigns to each standard bitableau $T = (A, B)$ of degree t a two-line array

$$\begin{pmatrix} u_1, u_2, \ldots, u_t \\ v_1, v_2, \ldots, v_t \end{pmatrix} \tag{6.4}$$

$$\begin{array}{|c|c|c|}\hline 1&2&4\\ \hline 3&5& \\ \cline{1-2} 4& & \\ \cline{1-1}\end{array} \quad\Longrightarrow\quad \begin{array}{|c|c|c|}\hline 1&3&4\\ \hline 4&5& \\ \cline{1-2}\end{array}$$

$$a_3 = 4, \;\; a_2 = 3, \;\; a_1 = 2$$

Figure 2

with $u_1 \leq u_2 \leq \cdots \leq u_t$ and $v_i \geq v_{i+1}$ if $u_i = u_{i+1}$. This assignment is defined as follows: starting with t we define recursively standard bitableaux and pairs of integers,

$$(A_t, B_t), (A_{t-1}, B_{t-1}), \ldots, (A_1, B_1) \quad \text{and} \quad u_t, v_t, u_{t-1}, v_{t-1}, \ldots, u_1, v_1.$$

We let $(A_t, B_t) = (A, B)$. Suppose (A_i, B_i) is already defined. Then let u_i the maximum of the elements in A_i, and s the maximal number with the property that u_i belongs the sth row of A_i. The tableau A_{i-1} is obtained from A_i by removing the element u_i form the sth row of A_i. Now we apply DELETE to B_i for the integer s, which we defined before and obtain v_i and B_{i-1} as output.

Applying this recursion to the bitableau (A_7, B_7) in Figure 1 we obtain $u_7 = 4$ and $s = 3$. Now DELETE applied to B_7 yields $v_7 = 3$ and the new bitableau in Figure 3:

$$\begin{array}{|c|c|c|}\hline 1&2&4\\ \hline 1&3& \\ \cline{1-2} 2& & \\ \cline{1-1}\end{array} \qquad \begin{array}{|c|c|c|}\hline 1&2&3\\ \hline 2&5& \\ \cline{1-2} 2& & \\ \cline{1-1}\end{array}$$

$$(A_6, B_6)$$

Figure 3

Proceeding in this way we obtain the two-line array

$$\begin{pmatrix} 1, 1, 2, 2, 3, 4, 4 \\ 2, 2, 5, 1, 2, 3, 3 \end{pmatrix}.$$

We may identify the two-line array (6.4) with the monomial $x_{u_1 v_1} \cdots x_{u_t v_t}$. Thus, since any standard monomial in $\delta \in K[X]$ can be identified with a bitableaux (A, B), the Knuth-Robinson-Schensted equivalence defines a map, denoted KRS, between the set of standard monomials of $K[X]$ to the set of monomials of $K[X]$.

Theorem 6.31. *The map* KRS *is a degree preserving bijection between the set of standard monomials and the set of monomials of* $K[X]$.

We still need another important result regarding this correspondence.

Theorem 6.32 (Schensted). *Let* $\delta = \delta_1 \cdots \delta_r$ *be a standard monomial and*

$$\mathrm{KRS}(\delta) = \begin{pmatrix} u_1, u_2, \ldots, u_t \\ v_1, v_2, \ldots, v_t \end{pmatrix}.$$

Then $\deg \delta_1$ *is the length of the longest increasing subsequence of* $v_1, v_2, \ldots, v_t$.

Now we are ready to compute the Gröbner basis of $I_t(X)$. For this purpose we fix a **diagonal monomial order**, that is, a monomial order $<$ which selects the diagonal of a minor as its initial term. In other words, if $\delta = [a_1 \cdots a_t | b_1 \cdots b_t]$, then

$$\mathtt{in}_<(\delta) = x_{a_1 b_1} x_{a_2 b_2} \cdots x_{a_t b_t}.$$

An example of a diagonal monomial order is the lexicographic order induced by

$$x_{11} > x_{12} > \cdots > x_{1n} > x_{21} > x_{22} > \cdots > x_{2n} > x_{31} > x_{32} > \cdots > x_{mn}.$$

Theorem 6.33. *For any diagonal monomial order the set of all* t*-minors of* X *is a Gröbner basis of* $I_t(X)$.

Proof. By Theorem 6.30, the standard monomials $\delta = \delta_1 \delta_2 \cdots \delta_r$ with $\deg \delta_1 \geq t$ form a K-basis of $I_t(X)$. Let us denote this set of standard monomials by D_t. Schensted's theorem implies that for each $\delta \in D_t$, the monomial $\mathrm{KRS}(\delta)$ contains as a factor the main diagonal of a t-minor of X. Therefore, for $\delta \in D_t$ there exists a t-minor σ such that $\mathtt{in}_<(\sigma) | \mathrm{KRS}(\delta)$. Thus, if J denotes the ideal generated by the initial monomials of the t-minors of X, we see that $\mathrm{KRS}(D_t) \subset J \subset \mathtt{in}_<(I_t(X))$.

For a subset $\mathcal{S} \subset K[X]$ consisting of homogeneous polynomials we denote by $K\mathcal{S}$ the K-vector space spanned by the elements of $\mathcal{S}$. Then $K\mathcal{S}$ is a graded K-vector space, whose ith graded component we denote by $(K\mathcal{S})_i$. Since $(KD_t)_i = I_t(X)_i$, Theorem 6.31 implies that

$$\begin{aligned} \dim_K I_t(X)_i &= \dim_K (K\,\mathrm{KRS}(D_t))_i \leq \dim_K J_i \\ &\leq \dim_K \mathtt{in}_<(I_t(X))_i = \dim_K I_t(X)_i \end{aligned}$$

for all i. Consequently, $\mathrm{Hilb}_{S/J}(t) = \mathrm{Hilb}_{S/\mathtt{in}_<(I_t(X))}(t)$. Thus the desired conclusion follows from Proposition 4.29. □

6.5.2. The initial complex of a determinantal ideal. In the previous subsection we have seen that for any diagonal monomial order $<$ the set of all t-minors of X is a Gröbner basis of $I_t(X)$. Hence for such a monomial order we have

$$\mathtt{in}_<(I_t(X)) = (\{x_{a_1b_1}x_{a_2b_2}\cdots x_{a_tb_t}\}_{1\leq a_1<a_2<\cdots<a_t\leq m,\ 1\leq b_1<b_2<\cdots<b_t\leq n}).$$

In particular, we see that $\mathtt{in}_<(I_t(X))$ is a squarefree initial ideal. Thus, identifying the variable x_{ij} with the point $(i,j)\in[m]\times[n]$, we may view $\mathtt{in}_<(I_t(X))$ as the Stanley–Reisner ideal of a simplicial complex on the vertex set $[m]\times[n]$. We denote this simplicial complex by Δ_t, and call it the **initial complex** of the determinantal ideal $I_t(X)$.

To better understand the nature of this complex we start with a very simple special case, namely the initial complex of the ideal of 2-minors of a generic $m\times 2$-matrix. In this case the vertex set is $[m]\times[2]$, and the minimal nonfaces are the subset of the form $\{(j,1),(k,2)\}$ with $1\leq j<k\leq m$.

Figure 4 shows the minimal nonface $\{(2,1),(4,2)\}$ in $[5]\times[2]$.

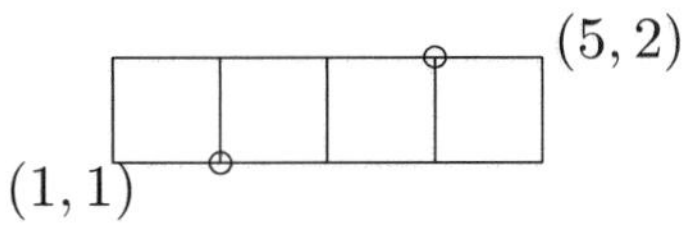

Figure 4

Thus, a facet of Δ_2 is a maximal subset of $[m]\times[2]$ which does not contain any pair of vertices in a position as indicated in Figure 4. In other words, if F is a facet of Δ_2 and $(i,2),(j,1)\in F$, then we must have $i\leq j$. It is then obvious that the facets of Δ_2 are the sets $F_k=\{(1,2),\ldots,(k,2),(k,1),\ldots,(m,1)\}$, $k=1,\ldots,m$.

One of the facets of Δ_2 in $[5]\times[2]$ is shown in Figure 5.

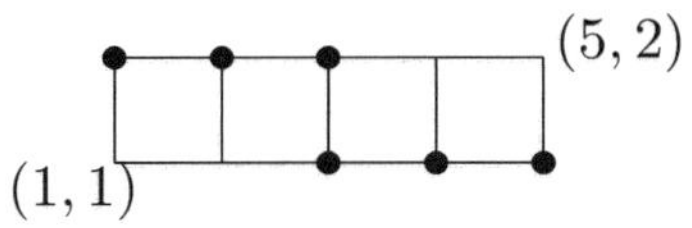

Figure 5

We define a partial order on $[m]\times[n]$ by setting

$$(i,j)\leq(k,l)\quad\text{if}\quad i\leq k\quad\text{and}\quad j\geq l. \tag{6.5}$$

With the partial order introduced, the facets of Δ_2 are just the maximal chains of the poset $[m] \times [2]$.

We will now study Δ_t for general t. Note that a t-antichain of $[m] \times [n]$ is a subset $\{(a_1, b_1), (a_2, b_2), \dots, (a_t, b_t)\} \in [m] \times [n]$ with $a_1 < a_2 < \cdots < a_t$ and $b_1 < b_2 < \cdots < b_t$. This shows that the set of t-antichains in $[m] \times [n]$ is in bijection with the minimal set of monomial generators of $\mathtt{in}_<(I_t(X))$, that is, with the minimal nonfaces of Δ_t.

Next we will identify the facets of Δ_t: let Z be a subset of $[m] \times [n]$. We denote by $\delta(Z)$ the set of all elements $(i, j) \in Z$ such that there exists no element $(k, l) \in Z$ with the property that $k < i$ and $l < j$. Obviously, $\delta(Z)$ is a chain of $[m] \times [n]$. An iterated application of the operation $Z \mapsto \delta(Z)$ yields a chain decomposition as follows:

$$\begin{aligned} Z_1 &= \delta(Z), \\ Z_i &= \delta(Z \setminus \bigcup_{j<i} Z_j), \quad \text{for} \quad i > 1. \end{aligned}$$

This decomposition of Z into disjoint sets of chains is obtained by the "light and shadow" procedure of Viennot, where the source of the light is in $(1, 1)$, see Figure 6.

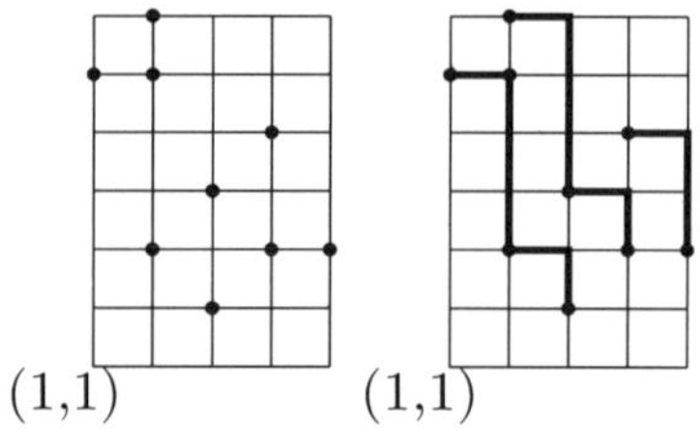

Figure 6

In the following we call a sequence of points $(u_1, v_1), \dots, (u_s, v_s)$ in $[m] \times [n]$ a **path** if $(u_i - u_{i-1}, v_i - v_{i-1})$ is equal to either $(1, 0)$ or $(0, -1)$ for all $i = 2, \dots, s$. Obviously, a path is a saturated chain of $[m] \times [n]$.

Theorem 6.34. *A subset $F \subset [m] \times [n]$ is a facet of Δ_t, if and only if F is the family of $t - 1$ nonintersecting paths from $P_r = (r, n)$ to $Q_r = (m, r)$ for $r = 1, \dots, t - 1$.*

Proof. We first observe that a subset $G \subset [n] \times [m]$ belongs to Δ_t if and only if G does not contain a t-antichain.

Suppose now that F is a family of $t - 1$ nonintersecting paths from $P_r = (r, n)$ to $Q_r = (m, r)$, $r = 1, \dots, t - 1$. Since an antichain intersects a chain in at most one point it follows that F does not contain a t-antichain, and hence it is a face of Δ_t. Suppose F is not a facet. Then there exists a vertex $P = (i,j) \notin F$ such that $F \cup \{P\}$ is a face of Δ_t.

Let

$$C = \{(k,l) \in [m] \times [n] \colon l - k \geq n - t + 1 \quad \text{or} \quad k - l \geq m - t + 1\}.$$

Figure 7 displays four nonintersecting paths in $[10] \times [7]$. The vertices of C belong to the upper left and lower right triangles indicated by the dotted lines.

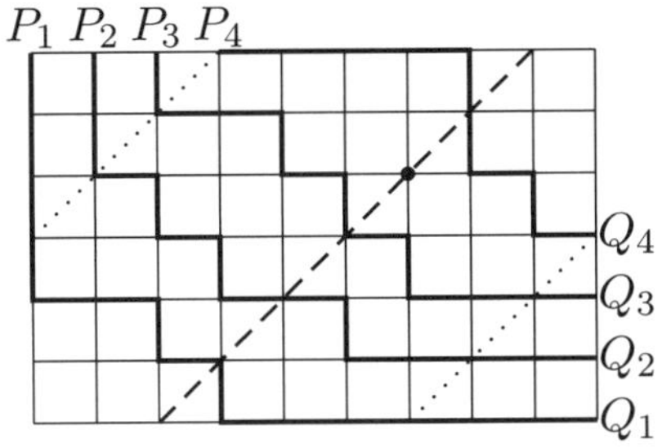

Figure 7

Since the vertices of C do not belong to any t-antichain, it follows that $C \subset F$. In particular, $P \notin C$. This implies that the set $\{(i+k, j+k) \colon\ k \in \mathbb{Z}\}$ intersects $F \cup \{P\}$ in t vertices; see Figure 7 where we have chosen $P = (7,5)$ and where the dashed line through P indicates the line $\{(7+k, 5+k) \colon\ k \in \mathbb{R}\}$ within $[10] \times [7]$. It follows that $(F \cup \{P\}) \cap \{(i+k, j+k) \colon\ k \in \mathbb{Z}\}$ is a t-antichain contained in $F \cup \{P\}$, a contradiction.

Conversely, let F be a facet of Δ_t, and let $F = \bigcup_r F_r$ be its decomposition into disjoint chains. We have $F_r = \emptyset$ for $r > t-1$, because otherwise F would contain a t-antichain, which is impossible. Moreover, since $C \subset F$, it follows that $P_r, Q_r \in F_r$ for $r = 1, \ldots, t-1$. Thus it remains to be shown that the chains F_r are saturated.

If $P_1 = (a_1, b_1)$ and $P_2 = (a_2, b_2)$, then we write $P_1 \prec P_2$, if $a_1 < a_2$ and $b_1 < b_2$. We note the following two facts:

(i) if $Q' \in F_k$, then for each $l < k$ there exists $P' \in F_l$ with $P' \prec Q'$;

(ii) if $Q' \in F_k$ and $P' \prec Q'$, then $P' \notin F_k$.

For the proof of (i) suppose that there exists an integer $l < k$ such that for no $P' \in F_l$ one has $P' \prec Q'$. Let l be the largest integer with this property. Then there exists $R \in F_{l+1}$ with $R \preceq Q'$ such that for no $P' \in F_l$ one has $P' \prec R$. Therefore there is no vertex $T \prec R$ with $T \in F \setminus \bigcup_{j=1}^{l-1} F_j$. By our definition of chain decompositions it follows that $R \in F_l$, a contradiction.

In order to prove (ii) assume that $P' \in F_k$. Then $P' \geq Q'$ or $P' \leq Q'$. Let $P' = (a,b)$ and $Q' = (c,d)$ with respect to the partial order defined in (6.5). In the first case we have $a \leq c$ and $b \geq d$, while in the second case $a \geq c$ and $b \leq d$. However, since $P' \prec Q'$ we have $a < c$ and $b < d$, a contradiction.

Now let $P, Q \in F_r$ with $Q > P$, and let $]P,Q[= \{R \in [m] \times [n] \colon Q > R > P\}$, where $<$ denotes the partial order defined in (6.5). It is enough to show that if $]P,Q[\neq \emptyset$, then $]P,Q[\cap F_r \neq \emptyset$.

Let $R \in]P,Q[$. If $R \notin F$, then

$$F \cup \{R\} = F_1 \cup \cdots \cup F_{r-1} \cup (F_r \cup \{R\}) \cup F_{r+1} \cup \cdots \cup F_{t-1}$$

is a decomposition of F into $t-1$ chains. Thus $F \cup \{R\}$ does not contain a t-antichain, and hence is a face of Δ_t, contradicting the fact that F is a facet of Δ_t. Thus we may assume that $R \in F$.

Consider first the case that P and Q are in horizontal position, that is, if $P = (a,b)$ and $Q = (c,d)$, then $b = d$ and $a < c$. Suppose $R \in F_s$ for some $s < r$. By (i), there exists $T \in F_s$ with $T \prec P$. Then $T \prec R$, contradicting (ii), since T and R both belong to F_s. On the other hand, if $R \in F_s$ for some $s > r$, then there exists $T \in F_r$ with $T \prec R$. It follows that $T \prec Q$, contradicting the fact that T and Q both belong to F_r. Similarly, one argues when P and Q are in vertical position.

Now consider the case that $c > a$ and $d < b$, and let $R = (c,b)$. Suppose $R \in F_s$ for some $s < r$. Let $T \in F_s$ with $T \prec P$. Then $T \prec R$, in contradiction to (ii), since $T, R \in F_s$. Finally, suppose that $R \in F_s$ for some $s > r$. Then there exists $T \in F_r$ with $T \prec R$. It follows from the choice of R that $P < T < Q$. Thus in all cases, we see that $]P,Q[\cap F_r \neq \emptyset$. $\square$

As a first consequence of Theorem 6.34 we obtain

Theorem 6.35. *The determinantal ring $K[X]/I_t(X)$ is an integral domain with*

$$\dim K[X]/I_t(X) = (m+n-t+1)(t-1).$$

Proof. By Corollary 4.30 and Theorem 5.9 one has

$$\begin{aligned} &\dim K[X]/I_t(X) = \dim K[X]/\operatorname{\mathtt{in}}_<(I_t(X)) \\ &= \dim \Delta_t + 1 = \max\{|F| \colon F \in \Delta_t\}. \end{aligned}$$

Since any path from P_r to Q_r has cardinality $m+n-(2r-1)$, it follows from Theorem 6.34 that all facets of Δ_t have cardinality

$$\sum_{r=1}^{t-1} (m+n-(2r-1)) = (t-1)(m+n-t+1).$$

This yields the desired dimension formula.

In order to see that $K[X]/I_t(X)$ is a domain, we proceed by induction on t. The assertion is trivial for $t = 1$. Now let $t > 1$; then x_{1n} does not belong to any diagonal of X. Let $<$ be a diagonal monomial order. Then Theorem 6.33 implies that x_{1n} does not divide any of the generators of $\mathtt{in}_<(I_t(X))$. Therefore x_{1n} is a nonzerodivisor on $K[X]/\mathtt{in}_<(I_t(X))$. Thus the subsequent lemma implies that x_{1n} is a nonzerodivisor on $K[X]/I_t(X)$ as well. It follows that the natural map $K[X]/I_t(X) \to (K[X]/I_t(X))_{x_{1n}}$ is injective. This implies that $K[X]/I_t(X)$ is a domain provided that the ring $(K[X]/I_t(X))_{x_{1n}}$ is a domain. That the latter ring is indeed a domain follows by induction, because

$$(K[X]/I_t(X))_{x_{1n}} \cong (K[Y]/I_{t-1}(Y))[x_{11}, \ldots, x_{1n}, x_{2n}, \ldots, x_{mn}][x_{1n}^{-1}],$$

where $Y = (y_{ij})_{\substack{i=2,\ldots,m \\ j=1,\ldots,n-1}}$ is a matrix of indeterminates. This isomorphism is induced by the isomorphism of K-algebras

$$K[X] \to K[Y][x_{11}, \ldots, x_{1n}, x_{2n}, \ldots, x_{mn}][x_{1n}^{-1}]$$

given by the substitutions

$$\begin{aligned} x_{ij} &\mapsto y_{ij} + x_{1j}x_{in}x_{1n}^{-1}, \quad \text{for} \quad i = 2, \ldots, m \quad \text{and} \quad j = 1, \ldots, n-1, \\ x_{ij} &\mapsto x_{ij}, \quad \text{for} \quad i = 1 \quad \text{or} \quad j = n. \end{aligned}$$

Indeed, this substitution maps X to a matrix Z from which one obtains the matrix X' with entries $x'_{1n} = x_{1n}$, $x'_{ij} = 0$ for $i = 1$ or $j = n$, and $x'_{ij} = y_{ij}$ for $i \geq 2$ and $j \leq n-1$ by clearing the last column and first row of Z with x_{1n} as a pivot element. It follows that for $I_t(Z)$, which is the image of $I_t(X)$ under this substitution, we have $I_t(Z) = x_{1n}I_{t-1}(X') = x_{1n}I_{t-1}(Y) = I_{t-1}(Y)$. □

Lemma 6.36. *Let K be a field, $S = K[x_1, \ldots, x_n]$ the polynomial ring over K in the indeterminates $x_1, \ldots, x_n$ and $I \subset S$ an ideal. Let $<$ be a monomial order on S and suppose that x_1 is a nonzerodivisor on $S/\mathtt{in}_<(I)$. Then x_1 is a nonzerodivisor on S/I.*

Proof. Let $x_1 f \in I$ for some $f \in S$, $f \neq 0$. Then $x_1 \mathtt{in}_<(f) \in \mathtt{in}_<(I)$. Our assumption implies that there exists $g \in I$ such that $\mathtt{in}_<(f) = \mathtt{in}_<(g)$. Thus we may choose $a \in K \setminus \{0\}$ such that $\mathtt{in}_<(h) < \mathtt{in}_<(f)$ for $h = f - ag$. Since $x_1 h \in I$, we now may assume, by using induction on the initial monomial, that $h \in I$. But then $f \in I$, as well. □

We have seen in the proof of Theorem 6.35 that Δ_t is a pure simplicial complex. We even have

Theorem 6.37. *The simplicial complex Δ_t is shellable.*

Proof. Let F and G be facets of Δ_t, and $F = \bigcup_{i=1}^{t-1} F_i$ and $G = \bigcup_{i=1}^{t-1} G_i$ their decomposition into disjoint paths. We set $F < G$ if for all i, G_i is on the upper right side of F_i. In other words, for all i and all $(a, b) \in G_i$ there exists $(c, d) \in F_i$ such that $a \geq c$ and $b \geq d$. We extend this partial order to a total order of the facets of Δ_t, and claim that this is a shelling order. To this end, we have to show that for $F < G$, there exists $P \in G \setminus F$ and a facet $H < G$ such that $G \setminus H = \{P\}$.

Since $F < G$, there exists some j such that F_j is not on the upper right side of G_j. Let i be the smallest integer with this property. Then there exists an upper corner $P = (a, b)$ of G_i which does not belong to F_i; see Figure 8. Here we call a point $P = (a, b)$ of a path R an upper corner of R, if both $(a - 1, b)$ and $(a, b - 1)$ belong to R.

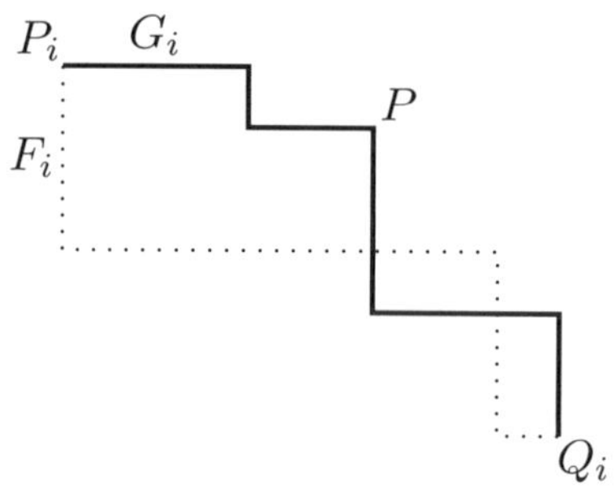

Figure 8

Let r be the largest integer such that $(a - k, b - k) \in G$ for $k = 1, \ldots, r$. Then $Q_k = (a - k, b - k) \in G_{i-k}$ for $k = 1, \ldots, r$. Set $Q_0 = P$. Then the desired facet H of Δ_t is obtained from G by replacing G_{i-k} by $G_{i-k} \setminus \{Q_k\} \cup \{Q_{k+1}\}$ for $k = 1, \ldots, r - 1$, and by letting $H_j = G_j$ for all other G_j. Figure 9 shows in an example how H arises from G. $\square$

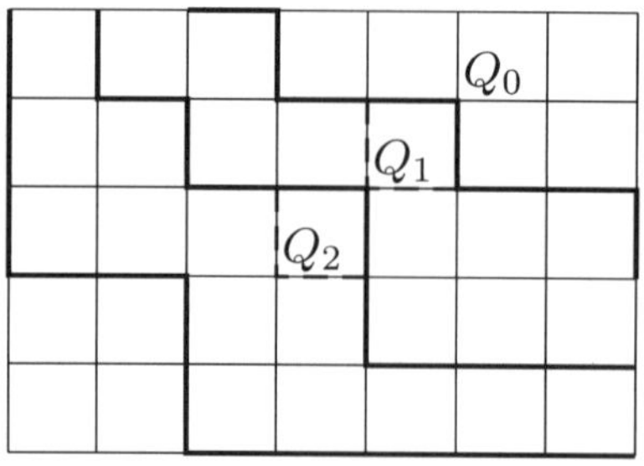

Figure 9

Theorem 6.37 together with Theorem 5.13 and Corollary 6.9 finally imply

Corollary 6.38. *The determinantal ring $K[X]/I_t(X)$ is Cohen–Macaulay.*

6.6. Sagbi bases and the coordinate ring of Grassmannians

6.6.1. Sagbi bases. Sagbi bases, introduced by Robbiano and Sweedler [**RS90**] and independently by Kapur and Madlener [**KM89**], are used to study subalgebras of polynomial rings. The philosophy is the same as that for Gröbner bases. By passing from a given subalgebra A of the polynomial ring $S = K[x_1, \dots, x_n]$ to the so-called initial algebra of A, one obtains an algebra $\mathtt{in}_<(A)$ generated over K by monomials of S. Many good properties of $\mathtt{in}_<(A)$, for example, being Cohen–Macaulay, are inherited by A. Since the structure of $\mathtt{in}_<(A)$ is usually much simpler than that of A, the study of the initial algebra of A provides a useful technique for the study of A itself.

The terminology "Sagbi" is the acronym for "**S**ubalgebra **a**nalog to **G**röbner **b**ases for **i**deals".

Definition 6.39. *Let K be a field and A a K-subalgebra of the polynomial ring $S = K[x_1, \dots, x_n]$. Given a monomial order $<$ on S we let $\mathtt{in}_<(A)$ be the K-subalgebra of S generated over K by all monomials $\mathtt{in}_<(f)$ with $f \in A$. The algebra $\mathtt{in}_<(A)$ is called the* ***initial algebra*** *of A with respect to the monomial order $<$.*

A set $\mathcal{S} \subset A$ is called a ***Sagbi basis*** *of A with respect to $<$, if the elements $\mathtt{in}_<(f)$ with $f \in \mathcal{S}$ generate the K-algebra $\mathtt{in}_<(A)$.*

In the following we will always assume that A is a finitely generated K-algebra. But, in general, this does *not* imply that $\mathtt{in}_<(A)$ is finitely generated, as the following example shows: let A be the subalgebra of $K[x,y]$ generated by $f_1 = x + y$, $f_2 = xy$ and $f_3 = xy^2$. Let $<$ be a monomial order with $x > y$. We show by induction on i that for all $i \geq 0$ the monomials xy^i belong to $\mathtt{in}_<(A)$. This is obviously the case for $i = 0, 1, 2$. Suppose now that $i > 2$, and that $xy^{i-1} \in \mathtt{in}_<(A)$. Then $xy^i = f_1(xy^{i-1}) - f_2(xy^{i-2}) \in \mathtt{in}_<(A)$. (The argument actually show that $xy^i \in A$ for $i > 0$.) Since $\mathtt{in}_<(A)$ is a monomial subalgebra of $K[x,y]$ containing for all $i \geq 0$ the monomial xy^i but no pure power of y, it follows that $\mathtt{in}_<(A)$ is not finitely generated; see Problem 6.8.

On the other hand, if $\mathtt{in}_<(A)$ is finitely generated, then so is A, as follows from the next result.

Proposition 6.40. *Assume that $\mathtt{in}_<(A) = K[\mathtt{in}_<(f_1), \dots, \mathtt{in}_<(f_m)]$. Then $A = K[f_1, \dots, f_m]$.*

Proof. Let $B = K[f_1, \dots, f_m]$ and assume that $B \neq A$. Let $f \in A \setminus B$ with smallest initial monomial. Since $\mathtt{in}_<(f) \in \mathtt{in}_<(A)$ there exist integers $a_i \geq 0$ and $c \in K$, $c \neq 0$, such that $\mathtt{in}_<(f) = c\,\mathtt{in}_<(f_1)^{a_1} \cdots \mathtt{in}_<(f_m)^{a_m}$. It follows that $g = f - cf_1^{a_1} \cdots f_m^{a_m} \in A$ with $\mathtt{in}_<(g) < \mathtt{in}_<(f)$. Thus we conclude that $g \in B$. But then $f \in B$ as well, a contradiction. □

In the case that A is generated by homogeneous polynomials, then A inherits a natural grading from S by setting $A_i = A \cap S_i$ for all i. In this situation we have

Proposition 6.41. *Let $A = K[f_1, \dots, f_m]$ be a K-subalgebra of S with the property that each $f_i \in S$ is a homogeneous polynomial. Then*

$$\mathrm{Hilb}_A(t) = \mathrm{Hilb}_{\mathtt{in}_<(A)}(t).$$

In particular, if $\mathtt{in}_<(A)$ is finitely generated, then A and $\mathtt{in}_<(A)$ have the same Krull dimension.

Proof. Given an integer i, and homogeneous polynomials $g_1, \dots, g_r$ with leading coefficient 1 such that $\mathtt{in}_<(g_1), \dots, \mathtt{in}_<(g_r)$ is a K-basis of $\mathtt{in}_<(A)_i$, then $g_1, \dots, g_r$ is a K-basis of A_i. Indeed, let $g \in A_i$, $g \neq 0$. Then there exists $c \in K$ such that either $g - cg_j = 0$ or $\mathtt{in}_<(g - cg_j) < \mathtt{in}_<(g)$. In the first case, we are done, and in the second case we may assume by an obvious induction argument that $g - cg_j \in \sum_{k=1}^r Kg_k$. Thus, again, we obtain the desired conclusion.

These considerations show that $\mathrm{Hilb}_A(t) = \mathrm{Hilb}_{\mathtt{in}_<(A)}(t)$. By [**BH98**, Proposition 4.4.1] the Krull dimension of a positively graded K-algebra is the pole order of its Hilbert series at $t = 1$. This implies the second assertion of the proposition. □

Remark 6.42. Assume that in the situation of Proposition 6.41, $\mathtt{in}_<(A)$ is generated in degree d, then A is generated in degree d, too. Thus we may consider both algebras to be standard graded, in which case it also follows directly from our definition of the Krull dimension of a graded module in Subsection 4.4.6 that $\dim A = \dim \mathtt{in}_<(A)$.

A deformation argument as it is used in the proof of Theorem 6.8 shows that $\operatorname{depth} \mathtt{in}_<(A) \leq \operatorname{depth} A$. In particular, if $\mathtt{in}_<(A)$ is Cohen–Macaulay, then so is A. For details we refer the reader to [**BC03**, Theorem 3.16].

The following criterion for Sagbi bases is the analogue to the Buchberger criterion and a variation of the criterion due to Robbiano and Sweedler [**RS90**].

Theorem 6.43. *Let $<$ be a monomial order on S, $f_1 \dots, f_m$ polynomials in S with leading coefficient 1 and $A = K[f_1, \dots, f_m]$ the K-subalgebra of S generated by $f_1 \dots, f_m$. Let $\varphi\colon R = K[y_1, \dots, y_m] \to A$ be a presentation* *of A with $\varphi(y_i) = f_i$ for $i = 1, \dots, m$, $J = \operatorname{Ker}(\varphi)$ the presentation ideal of A. Furthermore, let $B = K[\mathrm{in}_<(f_1), \dots, \mathrm{in}_<(f_m)]$ and $J_0 = \operatorname{Ker}(\psi)$ where $\psi\colon R = K[y_1, \dots, y_m] \to B$ is the K-algebra homomorphism with $\psi(y_i) = \mathrm{in}_<(f_i)$ for $i = 1, \dots, m$.*

Let $\mathbf{y}^{\mathbf{a}_1} - \mathbf{y}^{\mathbf{b}_1}, \dots, \mathbf{y}^{\mathbf{a}_r} - \mathbf{y}^{\mathbf{b}_r}$ be a system of binomial generators of the toric ideal J_0. Then $f_1, \dots, f_m$ is a Sagbi basis of A, if and only if the relations $\mathbf{y}^{\mathbf{a}_1} - \mathbf{y}^{\mathbf{b}_1}, \dots, \mathbf{y}^{\mathbf{a}_r} - \mathbf{y}^{\mathbf{b}_r}$ can be lifted to relations of A, that is, for each j there exist elements $c_{\mathbf{a}}^{(j)} \in K$ such that

$$\mathbf{f}^{\mathbf{a}_j} - \mathbf{f}^{\mathbf{b}_j} = \sum_{\mathbf{a}} c_{\mathbf{a}}^{(j)} \mathbf{f}^{\mathbf{a}} \quad \text{with} \quad \mathrm{in}_<(\mathbf{f}^{\mathbf{a}}) < \mathrm{in}_<(\mathbf{f}^{\mathbf{a}_j}) \quad \text{for all} \quad \mathbf{a}, \tag{6.6}$$

where $\mathbf{f}^{\mathbf{a}} = f_1^{a_1} \cdots f_m^{a_m}$ for $\mathbf{a} = (a_1, \dots, a_m)$.

If the equivalent conditions hold, then the polynomials

$$G_j(y_1, \dots, y_m) = \mathbf{y}^{\mathbf{a}_j} - \mathbf{y}^{\mathbf{b}_j} - \sum_{\mathbf{a}} c_{\mathbf{a}}^{(j)} \mathbf{y}^{\mathbf{a}}, \quad j = 1, \dots, r,$$

generate J.

Proof. Suppose that the given set of monomial generators of J_0 can be lifted. We first show that then any other binomial $\mathbf{y}^{\mathbf{c}} - \mathbf{y}^{\mathbf{d}}$ in J_0 can be lifted. To this end we give R a $\mathbb{Z}^n$-graded structure by setting $\deg y_j = \mathbf{c}_j$ where $\mathrm{in}_<(f_j) = \mathbf{x}^{\mathbf{c}_j}$. With this grading of R, the ideal J_0 is $\mathbb{Z}^n$-graded. Thus $\mathbf{y}^{\mathbf{c}} - \mathbf{y}^{\mathbf{d}}$ is a K-linear combination of binomials of the form $\mathbf{y}^{\mathbf{g}}(\mathbf{y}^{\mathbf{a}_j} - \mathbf{y}^{\mathbf{b}_j})$ with $\deg \mathbf{y}^{\mathbf{c}} = \deg \mathbf{y}^{\mathbf{g}}\mathbf{y}^{\mathbf{a}_j}$. This implies that $\mathrm{in}_<(\mathbf{f}^{\mathbf{c}}) = \mathrm{in}_<(\mathbf{f}^{\mathbf{g}})\,\mathrm{in}_<(\mathbf{f}^{\mathbf{a}_j})$. Since $\mathbf{y}^{\mathbf{a}_j} - \mathbf{y}^{\mathbf{b}_j}$ can be lifted we have

$$\mathbf{f}^{\mathbf{a}_j} - \mathbf{f}^{\mathbf{b}_j} = \sum_{\mathbf{a}} c_{\mathbf{a}}^{(j)} \mathbf{f}^{\mathbf{a}} \quad \text{with} \quad \mathrm{in}_<(\mathbf{f}^{\mathbf{a}}) < \mathrm{in}_<(\mathbf{f}^{\mathbf{a}_j}) \quad \text{for all} \quad \mathbf{a}.$$

It follows that $\mathbf{y}^{\mathbf{g}}(\mathbf{y}^{\mathbf{a}_j} - \mathbf{y}^{\mathbf{b}_j})$ is liftable, since

$$\mathbf{f}^{\mathbf{g}}(\mathbf{f}^{\mathbf{a}_j} - \mathbf{f}^{\mathbf{b}_j}) = \sum_{\mathbf{a}} c_{\mathbf{a}}^{(j)} \mathbf{f}^{\mathbf{g}}\mathbf{f}^{\mathbf{a}} \text{ with } \mathrm{in}_<(\mathbf{f}^{\mathbf{g}}\mathbf{f}^{\mathbf{a}}) < \mathrm{in}_<(\mathbf{f}^{\mathbf{g}})\,\mathrm{in}_<(\mathbf{f}^{\mathbf{a}_j}) = \mathrm{in}_<(\mathbf{f}^{\mathbf{c}})$$

for all $\mathbf{a}$.

Now since $\mathbf{f}^{\mathbf{c}} - \mathbf{f}^{\mathbf{d}}$ is a K-linear combination of elements of the form $\mathbf{f}^{\mathbf{g}}(\mathbf{f}^{\mathbf{a}_j} - \mathbf{f}^{\mathbf{b}_j})$ and since each of them is a K-linear combination of monomials in the f_i whose initial monomial is less than $\mathrm{in}_<(\mathbf{f}^{\mathbf{c}})$, it follows that $\mathbf{f}^{\mathbf{c}} - \mathbf{f}^{\mathbf{d}}$ can be expressed in the same way. In other words, $\mathbf{y}^{\mathbf{c}} - \mathbf{y}^{\mathbf{d}}$ is liftable.

Now we are going to show that $B = \mathrm{in}_<(A)$. To this end let $h \in A$, $h \neq 0$. We have to show that $\mathrm{in}_<(h) \in B$. Since $h \in A$ there exists a presentation $h = \sum_{\mathbf{a}} d_{\mathbf{a}} \mathbf{f}^{\mathbf{a}}$ with $d_{\mathbf{a}} \in K$. If $\mathrm{in}_<(\mathbf{f}^{\mathbf{a}}) \leq \mathrm{in}_<(h)$ for all $\mathbf{a}$ with $d_{\mathbf{a}} \neq 0$, then $\mathrm{in}_<(\mathbf{f}^{\mathbf{a}}) = \mathrm{in}_<(h)$ for some $\mathbf{a}$, and we are done. Otherwise

$$\max\{\mathrm{in}_<(\mathbf{f}^{\mathbf{a}})\colon\ d_{\mathbf{a}} \neq 0\} > \mathrm{in}_<(h).$$

Let $\mathbf{a}_1, \dots, \mathbf{a}_s$ be the exponents for which $\mathtt{in}_<(\mathbf{f}^\mathbf{a})$ is maximal. Then we have $\sum_{i=1}^s d_{\mathbf{a}_i} = 0$, and hence

$$\sum_{i=1}^s d_{\mathbf{a}_i}\mathbf{f}^{\mathbf{a}_i} = \sum_{i=2}^s d_{\mathbf{a}_i}(\mathbf{f}^{\mathbf{a}_i} - \mathbf{f}^{\mathbf{a}_1}).$$

Since $\mathtt{in}_<(\mathbf{f}^{\mathbf{a}_1}) = \mathtt{in}_<(\mathbf{f}^{\mathbf{a}_i})$, we see that $\mathbf{y}^{\mathbf{a}_1} - \mathbf{y}^{\mathbf{a}_i} \in J_0$, and since all the binomials in J_0 can be lifted, it follows that $\sum_{i=1}^s d_{\mathbf{a}_i}\mathbf{f}^{\mathbf{a}_i}$ can be rewritten as a K-linear combination L of monomials $\mathbf{f}^\mathbf{b}$ in the f_j with

$$\mathtt{in}_<(\mathbf{f}^\mathbf{b}) < \max\{\mathtt{in}_<(\mathbf{f}^\mathbf{a})\colon\ d_\mathbf{a} \neq 0\}.$$

Hence if we substitute the sum $\sum_{i=1}^s d_{\mathbf{a}_i}\mathbf{f}^{\mathbf{a}_i}$, which is part of the sum $\sum_\mathbf{a} d_\mathbf{a}\mathbf{f}^\mathbf{a}$, by the linear combination L, we obtain a new presentation $h = \sum_\mathbf{a} d'_\mathbf{a}\mathbf{f}^\mathbf{a}$ of h with

$$\max\{\mathtt{in}_<(\mathbf{f}^\mathbf{a})\colon\ d'_\mathbf{a} \neq 0\} < \max\{\mathtt{in}_<(\mathbf{f}^\mathbf{a})\colon\ d_\mathbf{a} \neq 0\}.$$

An obvious induction argument completes the proof.

Conversely, assume that $B = \mathtt{in}_<(A)$ and let $\mathbf{y}^\mathbf{c} - \mathbf{y}^\mathbf{d} \in J_0$. We want to show that $\mathbf{y}^\mathbf{c} - \mathbf{y}^\mathbf{d}$ is liftable. Since $\mathtt{in}_<(\mathbf{f}^\mathbf{c}) = \mathtt{in}_<(\mathbf{f}^\mathbf{d})$ it follows that $\mathtt{in}_<(\mathbf{f}^\mathbf{c} - \mathbf{f}^\mathbf{d}) < \mathtt{in}_<(\mathbf{f}^\mathbf{c})$. Since $B = \mathtt{in}_<(A)$, there exists $\mathbf{a}_1 \in \mathbb{N}^m$ and $c_{\mathbf{a}_1} \in K$ such that

$$\mathtt{in}_<(\mathbf{f}^{\mathbf{a}_1}) = \mathtt{in}_<(\mathbf{f}^\mathbf{c} - \mathbf{f}^\mathbf{d}) \quad \text{and} \quad \mathtt{in}_<(\mathbf{f}^\mathbf{c} - \mathbf{f}^\mathbf{d} - c_{\mathbf{a}_1}\mathbf{f}^{\mathbf{a}_1}) < \mathtt{in}_<(\mathbf{f}^{\mathbf{a}_1}).$$

As before we find $\mathbf{a}_2$ and $c_{\mathbf{a}_2} \in K$ such that

$$\mathtt{in}_<(\mathbf{f}^\mathbf{c} - \mathbf{f}^\mathbf{d} - c_{\mathbf{a}_1}\mathbf{f}^{\mathbf{a}_1} - c_{\mathbf{a}_2}\mathbf{f}^{\mathbf{a}_2}) < \mathtt{in}_<(\mathbf{f}^{\mathbf{a}_2}).$$

This process must terminate and yields after a finite number of steps the desired lifting.

Finally, we show that the relations $G_j(y_1, \dots, y_m)$ generate J. To see this, let $H(y_1, \dots, y_m) \in J$ with $H(y_1, \dots, y_m) = \sum_\mathbf{a} c_\mathbf{a} y^\mathbf{a}$ be an arbitrary element. By using the fact that $H(f_1, \dots, f_m) = 0$, the arguments presented above which showed that $B = \mathtt{in}_<(A)$ if the given set of generators can be lifted, also show that modulo the relations $G_j(y_1, \dots, y_m)$ the sum $\sum_\mathbf{a} c_\mathbf{a} y^\mathbf{a}$ can be rewritten as $\sum_\mathbf{a} c'_\mathbf{a} y^\mathbf{a}$ such that

$$\max\{\mathtt{in}_<(\mathbf{f}^\mathbf{a})\colon\ c'_\mathbf{a} \neq 0\} < \max\{\mathtt{in}_<(\mathbf{f}^\mathbf{a})\colon\ c_\mathbf{a} \neq 0\}.$$

Thus arguing by induction the desired conclusion follows. □

6.6.2. The coordinate ring of Grassmannians. Let K be a field and $X = (x_{ij})_{\substack{i=1,\dots,m \\ j=1,\dots,n}}$ a matrix of indeterminates. We denote by $K[X]$ the polynomial ring over K with indeterminates x_{ij}, and by A the K-subalgebra of $K[X]$ generated by all maximal minors of X. Notice that A is the coordinate ring of Grassmannians of the m-dimensional vector K-subspaces of K^n.

In this section we want to study A by means of its initial algebra with respect to the lexicographic order induced by the order of the variables

$$x_{11} > x_{12} > \cdots > x_{1n} > x_{21} > x_{22} > \cdots > x_{m1} > x_{m2} > \cdots > x_{mn}.$$

It is convenient to denote the maximal minor of X with columns $1 \leq a_1 < a_2 < \cdots < a_m \leq n$ by $[a_1, \ldots, a_m]$. Then

$$\mathtt{in}_<[a_1, \ldots, a_m] = x_{1a_1}x_{2a_2}\cdots x_{ma_m}.$$

Let B be the K-algebra generated by the initial monomials of all the maximal minors of X. Our ultimate goal will be to show that $B = \mathtt{in}_<(A)$. On the set $\mathcal{L}$ of all maximal minors we define a partial order be setting

$$[a_1, \ldots, a_m] \leq [b_1, \ldots, b_m] \quad \Leftrightarrow \quad a_i \leq b_i \quad \text{for all} \quad i.$$

This partially ordered set admits meets and joins; in other words, for any two elements $\delta_1, \delta_2 \in \mathcal{L}$ there exists a unique smallest element $\geq \delta_1, \delta_2$, called the **join** and denoted by $\delta_1 \vee \delta_2$, and a unique largest element $\leq \delta_1, \delta_2$, called the **meet** and denoted by $\delta_1 \wedge \delta_2$. In fact, we have

$$[a_1, \ldots, a_m] \vee [b_1, \ldots, b_m] = [\max\{a_1, b_1\}, \max\{a_2, b_2\}, \ldots, \max\{a_m, b_m\}]$$

and

$$[a_1, \ldots, a_m] \wedge [b_1, \ldots, b_m] = [\min\{a_1, b_1\}, \min\{a_2, b_2\}, \ldots, \min\{a_m, b_m\}].$$

It is obvious that $\mathcal{L}$ is a distributive lattice with meets and joins as described.

An element δ in a distributive lattice is called **join irreducible**, if it is different from the unique smallest element in the lattice and cannot be written as the join of two elements of the lattice which are properly smaller that δ.

In our lattice $\mathcal{L}$ of maximal minors, the join irreducible elements can be easily identified.

Lemma 6.44. *The join irreducible elements of $\mathcal{L}$ are*

$$\delta_{ik} = [1, 2, \ldots, i, i+1+k, i+2+k, \ldots, m+k] \tag{6.7}$$

with $i = 0, 1, \ldots, m-1$ and $k = 1, \ldots, n-m$.

Proof. Given an integer $0 \leq i \leq m-1$, we say that a minor $[a_1, a_2, \ldots, a_m]$ has a gap at position i if $a_{i+1} - a_i > 1$ where we have set $a_0 = 0$. Notice that the minors listed in (6.7) are exactly those which have precisely one gap. We claim that if $[a_1, a_2, \ldots, a_m]$ has at least two gaps, say at position i and position j with $i < j$, then $[a_1, a_2, \ldots, a_m]$ is not join irreducible. Indeed, we have

$$\begin{aligned}[a_1, a_2, \ldots, a_m] &= [a_1, a_2, \ldots, a_i, a_{i+1} - 1, a_{i+2}, \ldots, a_m] \\ &\vee [a_1, a_2, \ldots, a_j, a_{j+1} - 1, a_{j+2}, \ldots, a_m].\end{aligned}$$

This shows that each join irreducible minor is one of the minors listed in (6.7). Since each of these minors is obviously join irreducible, the desired conclusion follows. □

Let P be the subposet of $\mathcal{L}$ consisting of all join irreducible elements. By Birkhoff's theorem the ideal lattice $\mathcal{I}(P)$ of P is isomorphic to $\mathcal{L}$. Figure 10 shows the poset P in the case that $m = 3$ and $n = 5$.

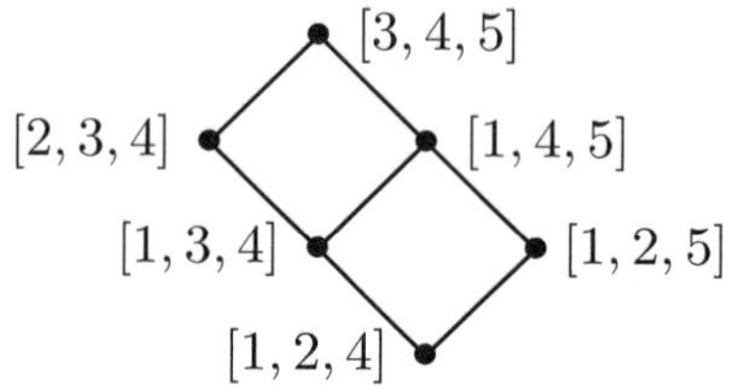

Figure 10

For the rest of this section we fix a diagonal monomial order $<$.

Theorem 6.45. *The K-algebra B generated over K by the initial monomials of the maximal minors of X is isomorphic to the Hibi ring attached to $\mathcal{L}$, that is, $B \cong K[\mathcal{L}]$, and* $\dim B = m(n-m)+1$.

Proof. Let T be the polynomial ring over K in the variables t_δ with $\delta \in \mathcal{L}$, and let $\psi\colon T \to B$ be the K-algebra homomorphism with $\psi(t_\delta) = \mathtt{in}_<(\delta)$. It is easily seen that the so-called Hibi relations

$$t_{\delta_1}t_{\delta_2} - t_{\delta_1\vee\delta_2}t_{\delta_1\wedge\delta_2}, \quad \delta_1, \delta_1 \in \mathcal{L} \tag{6.8}$$

belong to $J = \operatorname{Ker}\psi$. The Hibi relations are exactly the relations of the Hibi ring $K[\mathcal{L}]$; see [**HH10**, Theorem 10.1.3]. In particular, they generate a prime ideal $P_{\mathcal{L}}$. We will show that $J = P_{\mathcal{L}}$, thereby proving the theorem.

Since $P_{\mathcal{L}} \subset J$ and since both ideals are prime ideals, it will follow that $P_{\mathcal{L}} = J$, once we have shown that $\dim B = \dim K[\mathcal{L}]$. The dimension of $K[\mathcal{L}]$ is known to be 1 plus the cardinality of the underlying poset P (which is the subposet of join irreducible elements of $\mathcal{L}$); see [**Hi87**]. Thus it follows from Lemma 6.44 and a simple counting argument that $\dim K[\mathcal{L}] = m(n-m)+1$.

On the other hand, since B is an affine K-algebra we have $\dim B = \operatorname{tr deg} Q(B)/K$, where $Q(B)$ is the quotient field of B and "tr deg" denotes the transcendence degree of a field extension. Let

$$\mathcal{T} = \{\mathtt{in}_<(\delta)\colon\ \delta \in \mathcal{L},\ \delta \text{ is joint irreducible}\} \cup \{x_{11}x_{22}\cdots x_{mm}\}.$$

We now claim that $Q(B) = K(\mathcal{T})$, and that the set $\mathcal{T}$ is algebraically independent over K. This will then prove that indeed $\dim B = \dim K[\mathcal{L}] = m(n-m)+1$.

Let $\prec$ be a total order extending the partial order on $\mathcal{L}$, and let $\delta \in \mathcal{L}$. If $\delta \notin \mathcal{T}$, then there exist $\delta_1, \delta_2 \prec \delta$ such that $\delta = \delta_1 \vee \delta_2$. Thus

$$\mathtt{in}_<(\delta) = \mathtt{in}_<(\delta_1)\,\mathtt{in}_<(\delta_2)\,\mathtt{in}_<(\delta_1 \wedge \delta_2)^{-1}.$$

Since $\delta_1, \delta_2, \delta_1 \wedge \delta_2 \prec \delta$ we may assume by induction on the "size" of δ with respect to $\prec$ that $\mathtt{in}_<(\delta_1), \mathtt{in}_<(\delta_2), \mathtt{in}_<(\delta_1 \wedge \delta_2) \in K(\mathcal{T})$. Therefore, $\mathtt{in}_<(\delta) \in K(\mathcal{T})$. Hence we have shown that $\mathtt{in}_<(\delta) \in K(\mathcal{T})$ for all $\delta \in \mathcal{L}$. This shows that $Q(B) = K(\mathcal{T})$.

In order to see that the monomials in $\mathcal{T}$ are algebraically independent over K, we first notice that

$$\mathtt{in}_<(\delta_{ik}) = x_{11}x_{22}\cdots x_{ii}x_{i+1,i+1+k}\cdots x_{m,m+k}.$$

Thus, if we set $\mathbf{x} = x_{11}x_{22}\cdots x_{mm}$, then

$$\mathtt{in}_<(\delta_{ik})/\mathbf{x} = \mathbf{y}_{ik},$$

where

$$\mathbf{y}_{ik} = y_{i+1,i+1+k}y_{i+2,i+2+k}\cdots y_{m,m+k} \quad \text{and} \quad y_{jl} = x_{jl}/x_{jj}.$$

It follows that $K(\mathcal{T}) = K(\mathcal{Y})$, where

$$\mathcal{Y} = \{\mathbf{y}_{ik}\colon\ i = 0,\ldots,m-1, k = 1,\ldots,n-m\} \cup \{\mathbf{x}\}.$$

Next we observe that $\mathbf{y}_{ik}/\mathbf{y}_{i+1,k} = y_{i+1,i+1+k}$, so that the set

$$\mathcal{V} = \{y_{ik}\colon\ i = 1,\ldots,m, k = 1,\ldots,n-m\} \cup \{\mathbf{x}\}$$

generates $K(\mathcal{T})$ over K.

Consider the semigroup ring R generated over K by the set monomials $\mathcal{V}$. The exponent vectors in $\mathbb{Z}^{m(n-m)+1}$ of the elements in $\mathcal{V}$ are obviously linearly independent. This implies that R is a polynomial ring. In other words, the set $\mathcal{V}$ is algebraically independent over K. Since $|\mathcal{T}| = |\mathcal{V}|$, it follows that the set $\mathcal{T}$ is also algebraically independent over K, as desired. $\square$

Now we are in the position to prove

Theorem 6.46. *The maximal minors of X form a Sagbi basis of A. In other words, $B \cong \mathtt{in}_<(A)$.*

Proof. We apply Theorem 6.43 and have to show that the Hibi relations (6.8) can be lifted. To show this we use the fact the any product of two incomparable minors can be straightened. This means the following: let $\delta_1 = [a_1, a_2 \ldots, a_m]$ and $\delta_2 = [b_1, b_2, \ldots, b_m]$ be two incomparable minors (with respect to the partial order given in $\mathcal{L}$). Then $\delta_1\delta_2$ can be written as a linear combination ℓ of products of minors $[c_1, c_2, \ldots, c_m][d_1, d_2, \ldots, d_m]$ satisfying the following condition $(*)$ (see [**BH98**, Section 7.1 and 7.2]):

(1) $[c_1, c_2, \dots, c_m] \le [d_1, d_2, \dots, d_m]$ and
$[c_1, c_2, \dots, c_m] \le [\min\{a_1, b_1\}, \dots, \min\{a_m, b_m\}]$.

(2) The sequence $(c_1, c_2, \dots, c_m, d_1, d_2, \dots, d_m)$ arises from the sequence $(a_1, a_2 \dots, a_m, b_1, b_2, \dots, b_m)$ by a permutation.

For example, if $\delta_1 = [2, 3, 6]$ and $\delta_1 = [1, 4, 5]$, then $\delta_1\delta_2$ is a linear combination of the following products of minors:

$$[1, 2, 4][3, 5, 6], [1, 2, 3][4, 5, 6], [1, 3, 4][2, 5, 6], [1, 3, 5][2, 4, 6].$$

Among the potential summands in ℓ we have also $(\delta_1 \wedge \delta_2)(\delta_1 \vee \delta_2)$ which in our example is $[1, 3, 5][2, 4, 6]$. The crucial fact to observe is that any product of maximal minors $\sigma\tau$ satisfying $(*)$ which is different from $(\delta_1 \wedge \delta_2)(\delta_1 \vee \delta_2)$ has the property that $\mathtt{in}_<(\sigma\tau) < \mathtt{in}_<(\delta_1\delta_2) = \mathtt{in}_<((\delta_1 \wedge \delta_2)(\delta_1 \vee \delta_2))$. Indeed, let k be the integer such that

$$c_i = \min\{a_i, b_i\} \quad \text{and} \quad d_i = \max\{a_i, b_i\} \quad \text{for} \quad i = 1, \dots, k-1,$$

and $c_k \neq \min\{a_k, b_k\}$ or $d_k \neq \max\{a_k, b_k\}$. We first notice that due to (2) and the choice of k the sequence $(c_k, \dots, c_m, d_k, \dots, d_m)$ and the sequence $\mathcal{S} = (a_k, \dots, a_m, b_k, \dots, b_m)$ coincide up to a permutation of their elements. If $c_k \neq \min\{a_k, b_k\}$, then condition (1) implies that $c_k < \min\{a_k, b_k\}$ which is impossible because $c_k \in \mathcal{S}$. Assuming, without loss of generality, that $a_k < b_k$ we then have $c_k = a_k$ and $d_k \neq \max\{a_k, b_k\}$. Suppose that $d_k < \max\{a_k, b_k\}$. If $d_k = a_k$, we reach a contradiction since c_k and d_k are different elements of the sequence $\mathcal{S}$ while a_k is the unique smallest element in $\mathcal{S}$. On the other hand, if $d_k < \max\{a_k, b_k\}$ and $d_k \neq a_k$, there exists $j > k$ with $a_j < b_k$ such that $d_k = a_j$. It follows that $a_{k+1} \le a_j < b_k < b_{k+1}$, so that $\min\{a_{k+1}, b_{k+1}\} = a_{k+1}$. This implies that

$$a_k = c_k < c_{k+1} \le \min\{a_{k+1}, b_{k+1}\} = a_{k+1}.$$

It follows that $c_{k+1} = a_{k+1}$ since in between a_k and a_{k+1} there is no other element from the sequence $\mathcal{S}$, and c_{k+1} belongs to this sequence. Proceeding in the same way we get

$$a_{j-1} = c_{j-1} < c_j \le \min\{a_j, b_j\} = a_j.$$

Thus we deduce as before that $c_j = a_j$. Since we also have $d_k = a_j$ it follows that a_j appears twice in the sequence $c_k, \dots, c_m, d_k, \dots, d_m$ which is a permutation of the sequence $\mathcal{S}$. However, a_j appears exactly once in the sequence $(a_k, \dots, a_m)$ and cannot appear in $(b_k, \dots, b_m)$ since $a_j < b_k < \dots < b_m$.

We conclude that $d_k > \max\{a_k, b_k\}$. Now, we get

$$\mathtt{in}_<((\delta_1 \wedge \delta_2)(\delta_1 \vee \delta_2)) = x_{1,a_1}x_{1,b_1} \cdots x_{k-1,a_{k-1}}x_{k-1,b_{k-1}}x_{k,a_k}x_{k,b_k} \cdots ,$$

and

$$\mathtt{in}_<(\sigma\tau) = x_{1,a_1}x_{1,b_1}\cdots x_{k-1,a_{k-1}}x_{k-1,b_{k-1}}x_{k,a_k}x_{k,d_k}\cdots$$

with $d_k > b_k$. This shows that indeed $\mathtt{in}_<(\sigma\tau) < \mathtt{in}_<((\delta_1 \wedge \delta_2)(\delta_1 \vee \delta_2))$.

Since $\delta_1\delta_2 - \ell = 0$, the initial monomial of $\delta_1\delta_2$ must cancel against the initial monomial of a summand in ℓ. However, by what we have seen, this summand can only be $(\delta_1 \wedge \delta_2)(\delta_1 \vee \delta_2)$. Thus $(\delta_1 \wedge \delta_2)(\delta_1 \vee \delta_2)$ actually appears in ℓ with coefficient 1. It follows that

$$\delta_1\delta_2 - (\delta_1 \wedge \delta_2)(\delta_1 \vee \delta_2) = \ell',$$

where $\ell' = \ell - (\delta_1 \wedge \delta_2)(\delta_1 \vee \delta_2)$ is a linear combination of products of maximal minors $\sigma\tau$ with $\mathtt{in}_<(\sigma\tau) < \mathtt{in}_<(\delta_1\delta_2)$, as desired. □

Corollary 6.47. *The coordinate ring A of the Grassmannian of m-dimensional vector K-subspaces of K^n is a Gorenstein ring of dimension*

$$m(n-m)+1.$$

Proof. By Corollary 6.29, the Hibi ring B is Cohen–Macaulay. Moreover, $\dim B = m(n-m)+1$ as shown in Theorem 6.45. Thus by Proposition 6.41 and Remark 6.42 it follows that A is a Cohen–Macaulay ring of dimension $m(n-m)+1$. Hibi showed [**Hi87**] that a Hibi ring is Gorenstein, if and only if the underlying poset P is pure, that is, if all maximal chains in P have the same length. In our case all maximal chains of P have length $n-1$. Thus B is Gorenstein, and since B is standard graded it has a symmetric h-vector. Since $B \cong \mathtt{in}_<(A)$ and since by Proposition 6.41, $\mathtt{in}_<(A)$ and A have the same Hilbert function, we conclude that the h-vector of A is symmetric, as well. Since A is a domain, [**BH98**, Corollary 4.4.6] implies that A is Gorenstein. □

6.7. Binomial edge ideals

Classically one studies the edge ideal of a graph G on the vertex set $[n]$ which is generated by all the squarefree monomials $x_ix_j \in K[x_1,\dots,x_n]$ where $\{i,j\}$ is an edge of G. Edge ideals were introduced by Villarreal in [**V90**] and they have been widely studied since then.

In this section we consider a class of binomial ideals associated with graphs which is a natural generalization of the ideal of 2-minors of a $2\times n$-matrix of indeterminates. They were introduced in [**HHHKR10**] and appeared independently in [**O10**]. Our presentation follows [**HHHKR10**]. In simple terms, a binomial edge ideal is just an ideal generated by an arbitrary collection of 2-minors of a $2\times n$-matrix whose entries are all indeterminates.

Let G be a simple graph on the vertex set $[n]$, that is, without loops and multiple edges. Let K be a field and $S = K[x_1,\dots,x_n,y_1,\dots,y_n]$ the polynomial ring in $2n$ indeterminates. For any $1 \leq i < j \leq n$, we set $f_{ij} = x_i y_j - x_j y_i$. The **binomial edge ideal** associated with G is the ideal $J_G \subset S$ generated by all the binomials f_{ij} where $i < j$ and $\{i,j\}$ is an edge of G. In particular, the ideal of 2-minors of a $2 \times n$-matrix may be interpreted as the binomial edge ideal of the complete graph on $[n]$. Note that if G has an isolated vertex i, that is, there is no edge of G incident to i, then $J_G = J_{G'}$, where G' is the restriction of G to the vertex set $[n] \setminus \{i\}$.

In this subsection we are concerned with the computation of the Gröbner basis for binomial edge ideals. For this purpose we define the following notion.

Definition 6.48. *Let G be a simple graph G on $[n]$ and let i and j be two vertices of G with $i < j$. A path $i = i_0, i_1, \ldots, i_r = j$ from i to j is called **admissible** if the following conditions are fulfilled:*

(i) *$i_k \neq i_\ell$ for $k \neq \ell$;*

(ii) *for each $k = 1, \ldots, r-1$ one has either $i_k < i$ or $i_k > j$;*

(iii) *for any proper subset $\{j_1, \ldots, j_s\}$ of $\{i_1, \ldots, i_{r-1}\}$, the sequence $i, j_1, \ldots, j_s, j$ is not a path in G.*

With an admissible path

$$\pi : i = i_0, i_1, \ldots, i_r = j$$

from i to j where $i < j$, we associate the monomial

$$u_\pi = \Big(\prod_{i_k > j} x_{i_k}\Big)\Big(\prod_{i_\ell < i} y_{i_\ell}\Big).$$

In the next theorem we characterize the Gröbner basis of J_G with respect to the lexicographic order on S with $x_1 > \cdots > x_n > y_1 > \cdots > y_n$.

Theorem 6.49. *The set of binomials*

$$\mathcal{G} = \bigcup_{i<j} \{u_\pi f_{ij} : \pi \text{ is an admissible path from } i \text{ to } j\}$$

is a reduced Gröbner basis of J_G with respect to lexicographic order on S.

Proof. We organize this proof as follows: First we prove that $\mathcal{G} \subset J_G$. Then, since $\mathcal{G}$ is a system of generators, we show that $\mathcal{G}$ is a Gröbner basis of J_G by using Buchberger's criterion. Finally, in the last step, it is proved that $\mathcal{G}$ is a reduced Gröbner basis.

First Step. We show that, for each admissible path π from i to j, where $i < j$, the binomial $u_\pi f_{ij}$ belongs to J_G. Let $\pi : i = i_0, i_1, \ldots, i_{r-1}, i_r = j$ be an admissible path in G. We proceed by induction on r. Clearly, the assertion is true if $r = 1$. Let $r > 1$ and $A = \{i_k : i_k < i\}$ and $B = \{i_\ell : i_\ell > j\}$. One has either $A \neq \emptyset$ or $B \neq \emptyset$. If $A \neq \emptyset$, then we set $i_{k_0} = \max A$. If $B \neq \emptyset$, then we set $i_{\ell_0} = \min B$.

Suppose $A \neq \emptyset$. It then follows that each of the paths

$$\pi_1 : i_{k_0}, i_{k_0-1}, \ldots, i_1, i_0 = i \text{ and } \pi_2 : i_{k_0}, i_{k_0+1}, \ldots, i_{r-1}, i_r = j$$

in G is admissible. Now, the induction hypothesis guarantees that each of $u_{\pi_1} f_{i_{k_0},i}$ and $u_{\pi_2} f_{i_{k_0},j}$ belongs to J_G. A routine computation says that the S-polynomial $S(u_{\pi_1} f_{i_{k_0},i}, u_{\pi_2} f_{i_{k_0},j})$ is equal to $u_\pi f_{ij}$. Hence $u_\pi f_{ij} \in J_G$, as desired.

When $B \neq \emptyset$, the same argument as in the case $A \neq \emptyset$ is valid.

Second Step. It will be proven that the set of those binomials $u_\pi f_{ij}$, where π is an admissible path from i to j, forms a Gröbner basis of J_G. In order to show this we apply Buchberger's criterion, that is, we show that all S-pairs $S(u_\pi f_{ij}, u_\sigma f_{k\ell})$, where $i < j$ and $k < \ell$, reduce to zero. For this we will consider different cases.

In the case that $i = k$ and $j = \ell$, one has $S(u_\pi f_{ij}, u_\sigma f_{k\ell}) = 0$.

In the case that $\{i, j\} \cap \{k, \ell\} = \emptyset$, or $i = \ell$, or $k = j$, the initial monomials $\mathtt{in}_<(f_{ij})$ and $\mathtt{in}_<(f_{k\ell})$ form a regular sequence. Hence the S-pair $S(u_\pi f_{ij}, u_\sigma f_{k\ell})$ reduces to zero, because of the following more general fact: let $f, g \in S$ such that $\mathtt{in}_<(f)$ and $\mathtt{in}_<(g)$ form a regular sequence and let u and v be any monomials. Then $S(uf, vg)$ reduces to zero; see Problem 2.17.

It remains to consider the cases that either $i = k$ and $j \neq \ell$ or $i \neq k$ and $j = \ell$. Suppose we are in the first case. (The second case can be proved similarly.) We must show that $S(u_\pi f_{ij}, u_\sigma f_{i\ell})$ reduces to zero. We may assume that $j < \ell$, and must find a standard expression for $S(u_\pi f_{ij}, u_\sigma f_{i\ell})$ whose remainder is equal to zero.

Let $\pi : i = i_0, i_1, \ldots, i_r = j$ and $\sigma\colon i = i'_0, i'_1, \ldots, i'_s = \ell$. Then there exist indices a and b such that

$$i_a = i'_b \quad \text{and} \quad \{i_{a+1}, \ldots, i_r\} \cap \{i'_{b+1}, \ldots, i'_s\} = \emptyset.$$

Consider the path

$$\tau : j = i_r, i_{r-1}, \ldots, i_{a+1}, i_a = i'_b, i'_{b+1}, \ldots, i'_{s-1}, i'_s = \ell$$

from j to ℓ. To simplify the notation we write this path as

$$\tau\colon j = j_0, j_1, \ldots, j_t = \ell.$$

Let

$$j_{t(1)} = \min\{\, j_c \,:\, j_c > j,\ c = 1, \ldots, t\,\}$$

and

$$j_{t(2)} = \min\{\, j_c \,:\, j_c > j,\ c = t(1) + 1, \ldots, t\,\}.$$

Continuing these procedures yield the integers

$$0 = t(0) < t(1) < \cdots < t(q-1) < t(q) = t.$$

It then follows that

$$j = j_{t(0)} < j_{t(1)} < \cdots < j_{t(q)-1} < j_{t(q)} = \ell$$

and, for each $1 \leq c \leq t$, the path

$$\tau_c : j_{t(c-1)}, j_{t(c-1)+1}, \ldots, j_{t(c)-1}, j_{t(c)}$$

is admissible.

The crucial point of the proof is to show that

$$S(u_\pi f_{ij}, u_\sigma f_{i\ell}) = \sum_{c=1}^{q} v_{\tau_c} u_{\tau_c} f_{j_{t(c-1)} j_{t(c)}}$$

is a standard expression of $S(u_\pi f_{ij}, u_\sigma f_{i\ell})$ whose remainder is equal to 0, where each v_{τ_c} is the monomial defined as follows: Let $w = y_i \operatorname{lcm}(u_\pi, u_\sigma)$. Thus $S(u_\pi f_{ij}, u_\sigma f_{i\ell}) = -w f_{j\ell}$. Then

(i) if $c = 1$, then

$$v_{\tau_1} = \frac{x_\ell w}{u_{\tau_1} x_{j_{t(1)}}};$$

(ii) if $1 < c < q$, then

$$v_{\tau_c} = \frac{x_j x_\ell w}{u_{\tau_c} x_{j_{t(c-1)}} x_{j_{t(c)}}};$$

(iii) if $c = q$, then

$$v_{\tau_q} = \frac{x_j w}{u_{\tau_q} x_{j_{t(q-1)}}}.$$

Thus we have to show that

$$w f_{j\ell} = \frac{w x_\ell}{x_{j_{t(1)}}} f_{j j_{t(1)}} + \sum_{c=2}^{q-1} \frac{w x_j x_\ell}{x_{j_{t(c-1)}} x_{j_{t(c)}}} f_{j_{t(c-1)} j_{t(c)}} + \frac{w x_j}{x_{j_{t(q-1)}}} f_{j_{t(q-1)} \ell}$$

is a standard expression of $w f_{j\ell}$ with remainder 0. In other words, we must prove that

$$(\sharp) \quad \begin{aligned} w(x_j y_\ell - x_\ell y_j) = & \ \tfrac{w x_\ell}{x_{j_{t(1)}}} (x_j y_{j_{t(1)}} - x_{j_{t(1)}} y_j) \\ + & \ \textstyle\sum_{c=2}^{q-1} \frac{w x_j x_\ell}{x_{j_{t(c-1)}} x_{j_{t(c)}}} (x_{j_{t(c-1)}} y_{j_{t(c)}} - x_{j_{t(c)}} y_{j_{t(c-1)}}) \\ + & \ \tfrac{w x_j}{x_{j_{t(q-1)}}} (x_{j_{t(q-1)}} y_\ell - x_\ell y_{j_{t(q-1)}}) \end{aligned}$$

is a standard expression of $w(x_j y_\ell - x_\ell y_j)$ with remainder 0.

Since

$$
\begin{aligned}
wx_jy_\ell = \frac{wx_j}{x_{j_{t(q-1)}}}x_{j_{t(q-1)}}y_\ell &> \frac{wx_jx_\ell}{x_{j_{t(q-2)}}x_{j_{t(q-1)}}}x_{j_{t(q-2)}}y_{j_{t(q-1)}} \\
&> \cdots > \frac{wx_jx_\ell}{x_{j_{t(1)}}x_{j_{t(2)}}}x_{j_{t(1)}}y_{j_{t(2)}} > \frac{wx_\ell}{x_{j_{t(1)}}}x_jy_{j_{t(1)}},
\end{aligned}
$$

it follows that, if the equality ($\sharp$) holds, then ($\sharp$) turns out to be a standard expression of $w(x_jy_\ell - x_\ell y_j)$ with remainder 0. If we rewrite ($\sharp$) as

$$
\begin{aligned}
w(x_jy_\ell - x_\ell y_j) &= w(x_jx_\ell\frac{y_{j_{t(1)}}}{x_{j_{t(1)}}} - x_\ell y_j) \\
&+ wx_jx_\ell\sum_{c=2}^{q-1}(\frac{y_{j_{t(c)}}}{x_{j_{t(c)}}} - \frac{y_{j_{t(c-1)}}}{x_{j_{t(c-1)}}}) \\
&+ w(x_jy_\ell - x_jx_\ell\frac{y_{j_{t(q-1)}}}{x_{j_{t(q-1)}}}),
\end{aligned}
$$

then clearly the equality holds.

Third Step. Finally, we show that the Gröbner basis $\mathcal{G}$ is reduced. Let $u_\pi f_{ij}$ and $u_\sigma f_{k\ell}$, where $i < j$ and $k < \ell$, belong to $\mathcal{G}$ with $u_\pi f_{ij} \neq u_\sigma f_{k\ell}$. Let $\pi : i = i_0, i_1, \ldots, i_r = j$ and $\sigma : k = k_0, k_1, \ldots, k_s = \ell$. Suppose that $u_\pi x_i y_j$ divides either $u_\sigma x_k y_\ell$ or $u_\sigma x_\ell y_k$. Then $\{i_0, i_1, \ldots, i_r\}$ is a proper subset of $\{k_0, k_1, \ldots, k_s\}$.

Let $i = k$ and $j = \ell$. Then $\{i_1, \ldots, i_{r-1}\}$ is a proper subset of the set $\{k_0, k_1, \ldots, k_s\}$ and $k, i_1, \ldots, i_{r-1}, \ell$ is an admissible path. This contradicts the fact that σ is an admissible path.

Let $i = k$ and $j \neq \ell$. Then y_j divides u_σ. Hence $j < k$. This contradicts $i < j$.

Let $\{i, j\} \cap \{k, \ell\} = \emptyset$. Then $x_i y_j$ divides u_σ. Hence $i > \ell$ and $j < k$. This contradicts $i < j$. ☐

Corollary 6.50. *Every binomial edge ideal is a radical ideal.*

Proof. The statement follows immediately by the above theorem and the next easy lemma. ☐

Lemma 6.51. *Let $I \subset K[x_1, \ldots, x_n]$ be a graded ideal which has a squarefree initial ideal with respect to some monomial order. Then I is a radical ideal.*

Proof. Let $<$ be a monomial order on S such that $\mathtt{in}_<(I)$ is a squarefree monomial ideal. We have to show that $\sqrt{I} = I$. Let us assume that $I \subsetneq \sqrt{I}$, and let us choose a polynomial $f \in \sqrt{I} \setminus I$ which has the smallest initial monomial with respect to $<$ among all the homogeneous polynomials in $\sqrt{I} \setminus I$. There exists an integer $m \geq 1$ such that $f^m \in I$, which implies that $\mathtt{in}_<(f^m) = (\mathtt{in}_<(f))^m \in \mathtt{in}_<(I)$. Then there exists w, a squarefree monomial generator of $\mathtt{in}_<(I)$, which divides $(\mathtt{in}_<(f))^m$. It follows that w divides $\mathtt{in}_<(f)$ as well. Let $g \in I$ be a homogeneous polynomial such that $\mathtt{in}_<(g) = w$. We may choose a suitable term cu with $c \in K$ and u a monomial in S, such that the polynomial $f' := f - ug$ has initial monomial $\mathtt{in}_<(f')$ strictly smaller than $\mathtt{in}_<(f)$. On the other hand, it is obvious that $f' \in \sqrt{I} \setminus I$, a contradiction to the choice of f. □

By applying Theorem 6.49 we may characterize the binomial edge ideals which have a quadratic Gröbner basis with respect to the lexicographic order.

Corollary 6.52. *Let G be a simple graph on the vertex set $[n]$ with the edge set $E(G)$, and let $<$ be the lexicographic order on S induced by $x_1 > \cdots > x_n > y_1 > \cdots > y_n$. Then the following conditions are equivalent:*

(a) *The generators f_{ij} of J_G form a quadratic Gröbner basis.*

(b) *For all edges $\{i,j\}$ and $\{k,\ell\}$ with $i<j$ and $k<\ell$, one has $\{j,\ell\} \in E(G)$ if $i=k$ and $\{i,k\} \in E(G)$ if $j=\ell$.*

6.8. Connectedness of contingency tables

6.8.1. Contingency tables and the χ^2-statistics. In statistics, a **contingency table** is often used to record and analyze the relation between two or more categorical variables. It displays the (multivariate) frequency distribution of the variables in a matrix format. For example, suppose one hundred individuals from a large population are randomly sampled to find out whether there is any correlation between the color of their hair and the color of their eyes. A contingency table as displayed in Figure 11 records the result of this sampling.

The sequence of row and column sums is called the **marginal distribution** of the contingency table.

The question is whether there is any correlation between eye and hair color. To answer this question we have to decide on some statistical model (which is our null hypothesis) and to test to what extent the given table fits this model.

In general a (2-dimensional) contingency table is an $m \times n$-matrix whose entries are called the **cell frequencies**. Say that our contingency table has cell frequencies a_{ij}, while our statistical model gives the expected cell frequencies e_{ij}. Then the χ^2-statistic of the contingency table is computed by the formula

$$\chi^2 = \sum_{i,j} \frac{(a_{ij} - e_{ij})^2}{e_{ij}}.$$

		HAIR COLORS			
		Blonde	Red	Black	Totals
EYE COLORS	Brown	3	4	20	27
	Green	14	18	8	40
	Blue	16	12	5	33
Totals		33	34	33	100

Figure 11

Under the hypothesis of independence and homogeneity of proportions one has $e_{ij} = r_i c_j / N$ where $r_i = \sum_j a_{ij}$ is the ith row sum, $c_j = \sum_i a_{ij}$ is the jth column sum and $N = \sum_i r_i = \sum_j c_j$ is the total number of samples. In our example we obtain $\chi^2 = 29.001$.

6.8.2. Random walks. Does the value of χ^2 tell us that the data given in the table of Figure 11 fits well our hypothesis of independence and homogeneity of proportions? One strategy to answering this question is to compare the χ^2-statistic of the given table with a large number of randomly selected contingency tables with the same marginal distribution. If only a rather low percentage (which is commonly fixed to be 5 %) of those randomly selected contingency tables has a greater χ^2 than that of the given table, the null hypothesis is rejected.

But how can we produce such random contingency tables with fixed marginal distribution? We start at the given table A and take random moves that do not change the marginal distribution. Each single move is given as follows: choose a pair of rows and a pair of columns at random, and modify A at the four entries where the selected rows and columns intersect by adding or subtracting 1 according to the following pattern of signs

$$\begin{matrix} + & - \\ - & + \end{matrix} \quad \text{or} \quad \begin{matrix} - & + \\ + & - \end{matrix}$$

with probability 1/2 each. In this way we obtain a random walk on the set of contingency tables with fixed marginal distribution.

If the move produces negative entries, discard it and continue by choosing a new pair of rows and columns. We will show below that each table with the same marginal distribution can be obtained from A in a finite number of moves. If A is a contingency table of shape $m \times n$, then the number of possible moves is $\binom{m}{2}\binom{n}{2}$, which is a rather big number. In practice one obtains a pretty good selection of randomly selected contingency tables with the same marginal distribution as that of A which allows us to test the significance of A, if we restrict the set $\mathcal{S}$ of possible moves. For example, one could allow only moves where the selected rows and columns are consecutive. Another natural choice for the set of possible moves (studied by Glonek [**G87**]) is to allow only moves which involve the first row and first column. We say that two contingency tables A and B are **connected** via $\mathcal{S}$, if B can be obtained from A by a finite number of moves. The question arises how to decide whether two contingency tables are connected.

By converting the rows of a contingency table of shape $m \times n$ to a vector, we may view it as an element in the set $\mathbb{N}^{m\times n}$ of nonnegative integer vectors. Then the connectedness problem can be rephrased and generalized as follows: let $\mathcal{B}$ be a subset of vectors of $\mathbb{Z}^n$. One defines the graph $G_{\mathcal{B}}$ whose vertex set is the set $\mathbb{N}^n$ of nonnegative integer vectors. Two vectors $\mathbf{a}$ and $\mathbf{c}$ in $\mathbb{N}^n$ are connected by an edge of $G_{\mathcal{B}}$ if $\mathbf{a} - \mathbf{c} \in \pm\mathcal{B}$. Then the problem we have to deal with is to describe the connected components of this graph. We say that $\mathbf{a}$ and $\mathbf{c}$ are **connected via** $\mathcal{B}$, if they belong to the same connected component of $G_{\mathcal{B}}$.

Now we arrived at a point where we can use commutative algebra. We fix a field K and define the binomial ideal

$$I_{\mathcal{B}} = (\mathbf{x}^{\mathbf{b}^+} - \mathbf{x}^{\mathbf{b}^-} : \mathbf{b} \in \mathcal{B}) \subset K[x_1, \dots, x_n],$$

where for a vector $\mathbf{a} \in \mathbb{Z}^n$, the vectors $\mathbf{a}^+, \mathbf{a}^- \in \mathbb{N}^n$ are the unique vectors with $\mathbf{a} = \mathbf{a}^+ - \mathbf{a}^-$.

The crucial observation is the following:

Theorem 6.53. *The nonnegative integer vectors* $\mathbf{a}$ *and* $\mathbf{c}$ *are connected via* $\mathcal{B}$ *if and only if* $\mathbf{x}^{\mathbf{a}} - \mathbf{x}^{\mathbf{c}} \in I_{\mathcal{B}}$.

Proof. Suppose first that $\mathbf{a}$ and $\mathbf{c}$ are connected via $\mathcal{B}$. Then there exist $\mathbf{b}_1, \dots, \mathbf{b}_m \in \pm\mathcal{B}$ such that

$$\mathbf{a} + \mathbf{b}_1 + \mathbf{b}_2 + \cdots + \mathbf{b}_i \in \mathbb{N}^n \quad \text{for all} \quad i = 1, \dots, m,$$

and

$$\mathbf{c} = \mathbf{a} + \mathbf{b}_1 + \mathbf{b}_2 + \cdots + \mathbf{b}_m.$$

We show by induction on m that $\mathbf{x}^{\mathbf{a}} - \mathbf{x}^{\mathbf{c}} \in I_{\mathcal{B}}$. Suppose that $m = 1$. Then $\mathbf{c} = \mathbf{a} + \mathbf{b}_1 \in \mathbb{N}^n$, which implies that $\operatorname{supp}(\mathbf{b}_1^+) \subset \operatorname{supp}(\mathbf{c})$. It follows that $\mathbf{d} := \mathbf{a} - \mathbf{b}_1^- = \mathbf{c} - \mathbf{b}_1^+ \in \mathbb{N}^n$, and hence

$$\mathbf{x}^{\mathbf{a}} - \mathbf{x}^{\mathbf{c}} = \mathbf{x}^{\mathbf{d}}(\mathbf{x}^{\mathbf{b}_1^-} - \mathbf{x}^{\mathbf{b}_1^+}) \in I_{\mathcal{B}}.$$

Now suppose $m > 1$ and that the assertion is true for $i < m$. Since $\mathbf{a} + \mathbf{b}_1$ and $\mathbf{c}$ are connected via $m-1$ edges of $G_{\mathcal{B}}$, our induction hypothesis implies that $\mathbf{x}^{\mathbf{a}+\mathbf{b}_1} - \mathbf{x}^{\mathbf{c}} \in I_{\mathcal{B}}$, and this implies that

$$\mathbf{x}^{\mathbf{a}} - \mathbf{x}^{\mathbf{c}} = (\mathbf{x}^{\mathbf{a}} - \mathbf{x}^{\mathbf{a}+\mathbf{b}_1}) + (\mathbf{x}^{\mathbf{a}+\mathbf{b}_1} - \mathbf{x}^{\mathbf{c}}) \in I_{\mathcal{B}}.$$

Conversely, suppose that $\mathbf{x}^{\mathbf{a}} - \mathbf{x}^{\mathbf{c}} \in I_{\mathcal{B}}$. Then there exist $\mathbf{b}_1, \ldots, \mathbf{b}_m \in \pm\mathcal{B}$ and monomials $\mathbf{x}^{\mathbf{d}_i}$ such that

$$\mathbf{x}^{\mathbf{a}} - \mathbf{x}^{\mathbf{c}} = \sum_{i=1}^{m} \mathbf{x}^{\mathbf{d}_i}(\mathbf{x}^{\mathbf{b}_i^+} - \mathbf{x}^{\mathbf{b}_i^-});$$

see Problem 5.8.

We show by induction on m that $\mathbf{a}$ is connected to $\mathbf{c}$ via $\mathcal{B}$. If $m = 1$, then $\mathbf{x}^{\mathbf{a}} - \mathbf{x}^{\mathbf{c}} = \mathbf{x}^{\mathbf{d}_1}(\mathbf{x}^{\mathbf{b}_1^+} - \mathbf{x}^{\mathbf{b}_1^-})$. Therefore, $\mathbf{a} = \mathbf{d}_1 + \mathbf{b}_1^+$ and $\mathbf{c} = \mathbf{d}_1 + \mathbf{b}_1^-$, so that

$$\mathbf{a} - \mathbf{b}_1 = \mathbf{d}_1 + \mathbf{b}_1^+ - \mathbf{b}_1 = \mathbf{d}_1 + \mathbf{b}_1^- = \mathbf{c},$$

which means that $\mathbf{a}$ and $\mathbf{c} = \mathbf{d}_1 + \mathbf{b}_1^-$ are connected by the edge $-\mathbf{b}_1$. Now let $m > 1$. Then there exists an integer i, say $i = 1$, such that $\mathbf{x}^{\mathbf{a}} = \mathbf{x}^{\mathbf{d}_1+\mathbf{b}_1^+}$. It follows that

$$\mathbf{x}^{\mathbf{d}_1+\mathbf{b}_1^-} - \mathbf{x}^{\mathbf{c}} = \sum_{i=2}^{m} \mathbf{x}^{\mathbf{d}_i}(\mathbf{x}^{\mathbf{b}_i^+} - \mathbf{x}^{\mathbf{b}_i^-}).$$

Hence our induction hypothesis implies that $\mathbf{d}_1 + \mathbf{b}_1^-$ and $\mathbf{c}$ are connected via $\mathcal{B}$. Since $\mathbf{a}$ and $\mathbf{d}_1 + \mathbf{b}_1^-$ are connected by the edge $-\mathbf{b}_1$, the desired conclusion follows. □

As a first consequence we obtain

Corollary 6.54. *Let $\mathcal{C}(m,n)$ be the set of all contingency tables of shape $m \times n$, and $\mathcal{S}$ the set of all possible moves. Then any two contingency tables in $\mathcal{C}(m,n)$ with the same marginal distribution are connected via $\mathcal{S}$.*

Proof. The ideal $I_{\mathcal{S}}$ corresponding to the set of moves $\mathcal{S}$ is just the ideal $I_2(X)$ of all 2-minors of the matrix $X = (x_{ij})_{\substack{i=1,\ldots,m \\ j=1,\ldots,n}}$. By Theorem 6.35 we know that $I_2(X)$ is a prime ideal. Hence it follows from Proposition 5.4 that $I_{\mathcal{S}}$ is a lattice ideal of a sublattice $L \subset \mathbb{Z}^{m\times n}$. The lattice is generated by the $m \times n$-matrices of integers which correspond to the generators of $I_{\mathcal{S}}$, which are the 2-minors of X. Thus we see that the generators of L are matrices whose row and column sums are zero. It follows that all matrices in L have row and column sums equal to zero. Conversely, any integer $m \times n$-matrix with row and column sums equal to zero belongs to L; see Problem 6.27.

Now let A and B be two contingency tables with same marginal distribution, and consider the monomial $\mathbf{x}^A - \mathbf{x}^B$. Since A and B have the same marginal distribution, it follows that $A - B$ is an $m \times n$-matrix of integers with row and column sums equal to zero. This implies that $\mathbf{x}^A - \mathbf{x}^B \in I_{\mathcal{S}}$, and the desired conclusion follows from Theorem 6.53. □

6.8.3. Contingency tables of shape $2{\times}n$. In Corollary 6.54 we have seen how powerful Theorem 6.53 is in helping to decide whether two contingency tables are connected via the set $\mathcal{S}$ of all possible moves. In general, however, if we only allow a restricted proper subset $\mathcal{T} \subset \mathcal{S}$ of moves, then we obtain an ideal $I_{\mathcal{T}}$ which is properly contained in $I_2(X)$ and whose structure is much less understood than that of $I_2(X)$. The ideal $I_{\mathcal{T}}$ may not be a prime ideal, and not even a radical ideal; see Problem 6.28. Nevertheless, for contingency tables of shape $2 \times n$ there is a full-fledged theory, due to the fact that in this case the ideal $I_{\mathcal{T}}$ is nothing but the edge ideal of a suitable graph which was studied in Section 6.7.

In general we are confronted with the following problem. Given a binomial ideal I and a binomial f, can we find feasible conditions in terms of the exponents appearing in f that guarantee that $f \in I$? The following strategy suggested in [**DES98**] may be successful in some cases. Write the given binomial ideal I as an intersection $I = \bigcap_{k=1}^{r} J_k$ of ideals J_k. Then $f \in I$ if and only if $f \in J_k$ for all k. This strategy is useful only if each of the ideals J_k has a simple structure, so that it is possible to describe the conditions that guarantee that f belongs to J_k. A natural choice for such an intersection is a primary decomposition of I. In the case that I is a radical ideal the natural choice for the ideals J_k are the minimal prime ideals of I. While the existence of a primary decomposition of an ideal in a Noetherian ring can easily be shown in a nonconstructive way by using the ascending chain condition for ideals, it is much harder to give an algorithm which yields a primary decomposition for a given ideal. There exist several such algorithms (see for example [**GTZ88**] and [**EHV92**]), which are implemented in various computer algebra systems. Here we are dealing with binomial ideals. In [**ES84**] it is shown that a binomial ideal has a primary decomposition in terms of binomial ideals, provided that the base field is algebraically closed. In this context, a binomial ideal is understood to contain binomials and monomials. In a special case described below we give such a primary decomposition.

The binomial edge ideal of a graph is a radical ideal, as we have seen in Corollary 6.50. Thus if we want to apply the above described strategy to study connectedness of contingency tables of shape $2{\times}n$ we are lead to determine the minimal prime ideals of binomial edge ideals. The description of the minimal prime ideals of these ideals is taken from [**HHHKR10**]. Other interesting cases where the minimal prime ideals of a class of binomial ideals is known in quite explicit terms can be found, for example, in the papers [**HS00**], [**HS04**] and [**HH11**].

Let K be a field and $R = K[x_1, \dots, x_n, y_1, \dots, y_n]$ the polynomial ring in $2n$ indeterminates. We consider ideals generated by a given set of maximal minors of the $2 \times n$-matrix

$$\begin{pmatrix} x_1 & x_2 & \dots & x_n \\ y_1 & y_2 & \dots & y_n \end{pmatrix}.$$

Thus these ideals are generated by polynomials of the form $f_{ij} = x_i y_j - x_j y_i$ with $1 \le i < j \le n$. Any such ideal determines in a natural way a unique simple graph G on the vertex set $[n]$ with $\{i, j\}$ an edge of G if and only if f_{ij} is the generator of the ideal. The ideal of 2-minors associated with the graph G is called the **binomial edge ideal** of G and denoted J_G; cf. Section 6.7

Let G be a simple graph on $[n]$. For each subset $S \subset [n]$ we define a prime ideal $P_S(G) \subset R$. Let $T = [n] \setminus S$, and let $G_1, \dots, G_{c(S)}$ be the connected component of G_T. Here G_T is the induced subgraph of G whose edges are exactly those edges $\{i, j\}$ of G for which $i, j \in T$. For each G_i we denote by $\tilde{G}_i$ the complete graph on the vertex set $V(G_i)$. We set

$$P_S(G) = (\bigcup_{i \in S} \{x_i, y_i\}, J_{\tilde{G}_1}, \dots, J_{\tilde{G}_{c(S)}}).$$

Obviously, $P_S(G)$ is a prime ideal. In fact, each $J_{\tilde{G}_i}$ is the ideal of 2-minors of a generic $2 \times n_j$-matrix with $n_j = |V(G_j)|$. Since all the ideals $J_{\tilde{G}_j}$, as well as the ideal $(\bigcup_{i \in S} \{x_i, y_i\})$ are prime ideals in pairwise different sets of variables, $P_S(G)$ is a prime ideal, too.

Lemma 6.55. *With the notation introduced we have* height $P_S(G) = |S| + (n - c(S))$.

Proof. We have

$$\begin{aligned} \text{height}\, P_S(G) &= \text{height}(\bigcup_{i \in S} \{x_i, y_i\}) + \sum_{j=1}^{c(S)} \text{height}\, J_{\tilde{G}_j} = 2|S| + \sum_{j=1}^{c(S)} (n_j - 1) \\ &= |S| + (|S| + \sum_{j=1}^{c(S)} n_j) - c(S) = |S| + (n - c(S)), \end{aligned}$$

as required. Here we used that the ideal of 2-minors of a $2 \times n_j$-matrix has height $n_j - 1$; cf. Theorem 6.35. □

In **[ES84]** Eisenbud and Sturmfels showed that all associated prime ideals of a binomial ideal are binomial ideals in the sense that each polynomial generator has at most two terms. In our particular case we have

Theorem 6.56. *Let G be a simple graph on the vertex set $[n]$. Then* $J_G = \bigcap_{S \subset [n]} P_S(G)$.

Proof. It is obvious that each of the prime ideals $P_S(G)$ contains J_G. We will show by induction on n that each minimal prime ideal containing J_G is of the form $P_S(G)$ for some $S \subset [n]$. Since by Corollary 6.50, J_G is a radical ideal, and since a radical ideal is the intersection of its minimal prime ideals, the assertion of the theorem will follow.

Let P be a minimal prime ideal of J_G. We first show that $x_i \in P$ if and only if $y_i \in P$. For this part of the proof we may assume that G is connected. Indeed, if $G_1, \ldots, G_r$ are the connected components of G, then each minimal prime ideal P of J_G is of the form $P_1 + \cdots + P_r$ where each P_i is a minimal prime ideal of J_{G_i}. Thus if each P_i has the expected form, then so does P. Let $T = \{x_i\colon i \in [n], x_i \in P, y_i \notin P\}$. We will show that $T = \emptyset$. This will then imply that if $x_i \in P$, then $y_i \in P$. By symmetry it then also follows that $y_i \in P$ implies $x_i \in P$, so that the final conclusion will be that $x_i \in P$ if and only $y_i \in P$.

We first observe that $T \neq \{x_1, \ldots, x_n\}$. Because otherwise we would have $J_G \subset J_{\tilde{G}} \subsetneq (x_1, \ldots, x_n) \subset P$, and P would not be a minimal prime ideal of J_G.

Suppose that $T \neq \emptyset$. Since $T \neq \{x_1, \ldots, x_n\}$, and since G is connected there exists $\{i, j\} \in E(G)$ such that $x_i \in T$ but $x_j \notin T$. Since $x_iy_j - x_jy_i \in J_G \subset P$, and since $x_i \in P$ it follows that $x_jy_i \in P$. Hence since P is a prime ideal, we have $x_j \in P$ or $y_i \in P$. By the definition of T the second case cannot happen, and so $x_j \in P$. Since $x_j \notin T$, it follows that $y_j \in P$.

Let G' be the induced subgraph of G with vertex set $[n] \setminus \{j\}$. Then

$$(J_{G'}, x_j, y_j) = (J_G, x_j, y_j) \subset P.$$

Thus $\bar{P} = P/(x_j, y_j)$ is a minimal prime ideal of $J_{G'}$ with $x_i \in \bar{P}$ but $y_i \notin \bar{P}$ for all $x_i \in T \subset \bar{P}$. By induction hypothesis, $\bar{P}$ is of the form $P_S(G')$ for some subset $S \subset [n] \setminus \{j\}$. This contradicts the fact that $T \neq \emptyset$.

Now let G be again an arbitrary simple graph. By what we have shown it follows that there exists a subset $S \subset [n]$ such that $P = (\bigcup_{i \in S}\{x_i, y_i\}, \bar{P})$ where $\bar{P}$ is a prime ideal containing no variables. Let G' be the graph $G_{[n]\setminus S}$. Then reduction modulo the ideal $(\bigcup_{i \in S}\{x_i, y_i\})$ shows that $\bar{P}$ is a binomial prime ideal $J_{G'}$ which contains no variables. Let $G_1, \ldots, G_c$ be the connected components of G'. We will show that $\bar{P} = (J_{\tilde{G}_1}, \ldots, J_{\tilde{G}_c})$. This then implies that $P = (\bigcup_{i \in S}\{x_i, y_i\}, J_{\tilde{G}_1}, \ldots, J_{\tilde{G}_c})$, as desired.

To simplify notation we may as well assume that P itself contains no variables and have to show that $P = (J_{\tilde{G}_1}, \ldots, J_{\tilde{G}_c})$, where $G_1, \ldots, G_c$ are the connected components of G. In order to prove this we claim that if i, j with $i < j$ is an edge of $\tilde{G}_k$ for some k, then $f_{ij} \in P$. From this it will then follow that $(J_{\tilde{G}_1}, \ldots, J_{\tilde{G}_c}) \subset P$. Since $(J_{\tilde{G}_1}, \ldots, J_{\tilde{G}_c})$ is a prime ideal containing J_G, and P is a minimal prime ideal containing J_G, we conclude that $P = (J_{\tilde{G}_1}, \dots, J_{\tilde{G}_c})$.

Let $i = i_0, i_1, \dots, i_r = j$ be a path in G_k from i to j. We proceed by induction on r to show that $f_{ij} \in P$. The assertion is trivial for $r = 1$. Suppose now that $r > 1$. Our induction hypothesis says that $f_{i_1 j} \in P$. On the other hand, one has $x_{i_1} f_{ij} = x_j f_{i i_1} + x_i f_{i_1 j}$. Thus $x_{i_1} f_{ij} \in P$. Since P is a prime ideal and since $x_{i_1} \notin P$, we see that $f_{ij} \in P$. $\square$

Now let G be an arbitrary simple graph. Which of the ideals $P_S(G)$ are minimal prime ideals of J_G? The following result helps to find them.

Proposition 6.57. *Let G be a simple graph on $[n]$, and let S and T be subsets of $[n]$. Let $G_1, \dots, G_s$ be the connected components of $G_{[n]\setminus S}$, and $H_1, \dots, H_t$ the connected components of $G_{[n]\setminus T}$. Then $P_T(G) \subset P_S(G)$, if and only if $T \subset S$ and for all $i = 1, \dots, t$ one has $V(H_i) \setminus S \subset V(G_j)$ for some j.*

Proof. For a subset $U \subset [n]$ we let L_U be the ideal generated by the variables $\{x_i, y_i\colon\ i \in U\}$. With this notation introduced we have $P_S(G) = (L_S, J_{\tilde{G}_1}, \dots, J_{\tilde{G}_s})$ and $P_T(G) = (L_T, J_{\tilde{H}_1}, \dots, J_{\tilde{H}_t})$. Hence it follows that $P_T(G) \subset P_S(G)$, if and only if

$$T \subset S \quad \text{and} \quad (L_S, J_{\tilde{H}_1}, \dots, J_{\tilde{H}_t}) \subset (L_S, J_{\tilde{G}_1}, \dots, J_{\tilde{G}_s}).$$

Observe that $(L_S, J_{\tilde{H}_1}, \dots, J_{\tilde{H}_t}) = (L_S, J_{\tilde{H}'_1}, \dots, J_{\tilde{H}'_t})$ where $H'_i = (H_i)_{[n]\setminus S}$. Therefore, $P_T(G) \subset P_S(G)$ if and only if

$$(L_S, J_{\tilde{H}'_1}, \dots, J_{\tilde{H}'_t}) \subset (L_S, J_{\tilde{G}_1}, \dots, J_{\tilde{G}_s})$$

which is the case if and only if $(J_{\tilde{H}'_1}, \dots, J_{\tilde{H}'_t}) \subset (J_{\tilde{G}_1}, \dots, J_{\tilde{G}_s})$, because the generators of the ideals $(J_{\tilde{H}'_1}, \dots, J_{\tilde{H}'_t})$ and $(J_{\tilde{G}_1}, \dots, J_{\tilde{G}_s})$ have no variables in common with the x_i and y_i for $i \in S$.

Since $V(H'_i) = V(H_i) \setminus S$, the assertion will follow once we have shown the following claim: let $A_1, \dots, A_s$ and $B_1, \dots, B_t$ be pairwise disjoint subsets of $[n]$. Then

$$(J_{\tilde{A}_1}, \dots, J_{\tilde{A}_s}) \subset (J_{\tilde{B}_1}, \dots, J_{\tilde{B}_t}),$$

if and only if for each $i = 1, \dots, s$ there exists a j such that $A_i \subset B_j$.

It is obvious that if the conditions on the A_i and B_j are satisfied, then we have the desired inclusion of the corresponding ideals.

Conversely, suppose that $(J_{\tilde{A}_1}, \dots, J_{\tilde{A}_s}) \subset (J_{\tilde{B}_1}, \dots, J_{\tilde{B}_t})$. Without loss of generality, we may assume that $\bigcup_{j=1}^t B_j = [n]$. Consider the surjective K-algebra homomorphism

$$\epsilon\colon R \to K[\{x_i, x_i z_1\}_{i \in B_1}, \dots, \{x_i, x_i z_t\}_{i \in B_t}] \subset K[x_1, \dots, x_n, z_1, \dots, z_t]$$

with $\epsilon(x_i) = x_i$ for all i and $\epsilon(y_i) = x_i z_j$ for $i \in B_j$ and $j = 1, \ldots, t$. Then

$$\operatorname{Ker}(\epsilon) = (J_{\tilde{B}_1}, \ldots, J_{\tilde{B}_t}).$$

Indeed, one may check that the set $\bigcup_{k=1}^{t}\{x_i y_j - x_j y_i \colon i < j,\, i, j \in B_k\}$ is a Gröbner basis of $\operatorname{Ker}(\epsilon)$ with respect to the lexicographic order induced by $x_1 > \cdots > x_n > y_1 > \cdots > y_n$.

Now fix one of the sets A_i and let $k \in A_i$. Then $k \in B_j$ for some k. We claim that $A_i \subset B_j$. Indeed, let $\ell \in A_i$ with $\ell \neq k$ and suppose that $\ell \in B_r$ with $r \neq j$. Since $x_k y_\ell - x_\ell y_k \in J_{\tilde{A}_i} \subset (J_{\tilde{B}_1}, \ldots, J_{\tilde{B}_t})$, it follows that $x_k y_\ell - x_\ell y_k \in \operatorname{Ker}(\epsilon)$, so that $0 = \epsilon(x_k y_\ell - x_\ell y_k) = x_k x_\ell z_j - x_k x_\ell z_r$, a contradiction. □

Let $G_1, \ldots, G_r$ be the connected components of G. Once we know the minimal prime ideals of J_{G_i} for each i the minimal prime ideals of J_G are known. Indeed, since the ideals J_{G_i} are ideals in different sets of variables, it follows that the minimal prime ideals of J_G are exactly the ideals $\sum_{i=t}^{r} P_i$ where each P_i is a minimal prime ideal of J_{G_i}.

The next result detects the minimal prime ideals of J_G when G is connected.

Corollary 6.58. *Let G be a connected simple graph on the vertex set $[n]$, and $S \subset [n]$. Then $P_S(G)$ is a minimal prime ideal of J_G if and only if $S = \emptyset$, or $S \neq \emptyset$ and for each $i \in S$ one has $c(S \setminus \{i\}) < c(S)$.*

Proof. Assume that $P_S(G)$ is a minimal prime ideal of J_G and fix $i \in S$. Let $G_1, \ldots, G_r$ be the connected components of $G_{[n]\setminus S}$. We distinguish several cases.

Suppose that there is no edge $\{i, j\}$ of G such that $j \in G_k$ for some k. Set $T = S \setminus \{i\}$. Then the connected components of $G_{[n]\setminus T}$ are $G_1, \ldots, G_r, \{i\}$. Thus $c(T) = c(S) + 1$. However, this case cannot happen, since Proposition 6.57 would imply that $P_T(G) \subset P_S(G)$.

Next suppose that there exists exactly one G_k, say G_1, for which there exists $j \in G_1$ such that $\{i, j\}$ is an edge of G. Then the connected components of $G_{[n]\setminus T}$ are $G_1', G_2, \ldots, G_r$ where $V(G_1') = V(G_1) \cup \{i\}$. Thus $c(T) = c(S)$. Again, this case cannot happen since Proposition 6.57 would imply that $P_T(G) \subset P_S(G)$.

It remains the case that there are at least two components, say $G_1, \ldots, G_k$, $k \geq 2$, and $j_\ell \in G_\ell$ for $\ell = 1, \ldots, k$ such that $\{i, j_\ell\}$ is an edge of G. Then the connected components of $G_{[n]\setminus T}$ are $G_1', G_{k+1}, \ldots, G_r$, where $V(G_1') = \bigcup_{\ell=1}^{k} V(G_\ell) \cup \{i\}$. Hence, in this case, $c(T) < c(S)$.

Conversely, suppose that $c(S \setminus \{i\}) < c(S)$ for all $i \in S$. We want to show that $P_S(G)$ is a minimal prime ideal of J_G. Suppose this is not the case. Then there exists a proper subset $T \subset S$ with $P_T(G) \subset P_S(G)$. We choose $i \in S \setminus T$. By assumption, we have $c(S \setminus \{i\}) < c(S)$. The discussion of the three cases above shows that we may assume that $G_1', G_{k+1}, \ldots, G_r$ are the components of $G_{([n]\setminus S)\cup\{i\}}$ where $V(G_1') = \bigcup_{\ell=1}^{k} V(G_\ell) \cup \{i\}$ and where $k \geq 2$. It follows that $G_{[n]\setminus T}$ has one connected component H which contains G_1'. Then $V(H) \setminus S$ contains the subsets $V(G_1)$ and $V(G_2)$. Hence $V(H) \setminus S$ is not contained in any $V(G_i)$. According to Proposition 6.57, this contradicts the assumption that $P_T(G) \subset P_S(G)$. □

In the terminology of graph theory, Corollary 6.58 says that if G is a connected graph, then $P_S(G)$ is a minimal prime ideal of J_G, if and only if each $i \in S$ is a **cut-point** of the graph $G_{([n]\setminus S)\cup\{i\}}$.

The following example demonstrates Corollary 6.58.

Example 6.59. Consider the path graph G of length 4 as shown in Figure 12. Then the only subsets $S \subset [4]$, besides the empty set, for which

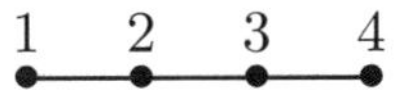

Figure 12

each $i \in S$ is a cut-point of the graph $G_{([4]\setminus S)\cup\{i\}}$, are the sets $S = \{2\}$ and $S = \{3\}$. Thus

$$J_G = I_2(X) \cap (x_2, y_2, x_3y_4 - x_4y_3) \cap (x_3, y_3, x_1y_2 - x_2y_1),$$

where

$$X = \begin{pmatrix} x_1 & x_2 & x_3 & x_4 \\ y_1 & y_2 & y_3 & y_4 \end{pmatrix}.$$

Next we analyze what it means that a binomial f belongs to a prime ideal of type $P_S(G)$. For that we need the following

Lemma 6.60. *Let $X_1 = (x_{ij}^{(1)}), \ldots, X_r = (x_{ij}^{(r)})$ be t-row matrices (with possibly different numbers of columns) in pairwise disjoint sets of variables. Then the binomial*

$$f = \prod_{k=1}^{r} \prod_{i,j} (x_{ij}^{(k)})^{a_{ij}^{(k)}} - \prod_{k=1}^{r} \prod_{i,j} (x_{ij}^{(k)})^{b_{ij}^{(k)}}$$

belongs to $I = I_2(X_1) + I_2(X_2) + \cdots + I_2(X_r)$ if and only if

$$\sum_j a_{ij}^{(k)} = \sum_j b_{ij}^{(k)} \quad \text{and} \quad \sum_i a_{ij}^{(k)} = \sum_i b_{ij}^{(k)} \quad \text{for all} \quad k.$$

Proof. The ideal I is a binomial prime ideal. Hence by Proposition 5.4, I is a lattice ideal. In fact, if n_k is the number of columns of X_k, then $I = I_L$ where

$$L \subset \mathbb{Z}^{t\times n_1} \times \mathbb{Z}^{t\times n_2} \times \cdots \times \mathbb{Z}^{t\times n_r}$$

with $L = L_1 \times L_2 \times \cdots \times L_r$, and $L_k \subset \mathbb{Z}^{t\times n_k}$ consists of all integer matrices $(c_{ij}^{(k)}) \subset \mathbb{Z}^{t\times n_k}$ with all row and column sums equal to zero. Here we identify row-wise the $t \times n_k$-matrix $(c_{ij}^{(k)})$ with the corresponding $t \times n_k$-vector.

We may assume that f is primitive. Then $f \in I$ if and only if

$$(a_{ij}^{(1)} - b_{ij}^{(1)}) \times (a_{ij}^{(2)} - b_{ij}^{(2)}) \times \cdots \times (a_{ij}^{(r)} - b_{ij}^{(r)}) \in L,$$

and this is the case if and only for all k, the matrix $(a_{ij}^{(k)} - b_{ij}^{(k)})$ has row and column sums equal to zero. This yields the desired conclusion. □

Let $A = (a_{ij})_{\substack{i=1,2\\ j=1,\ldots,n}}$ be a $2\times n$-matrix. The set $\operatorname{supp}(A) = \{(i,j)\colon\ a_{ij} \neq 0\}$ is called the **support** of A. If $T \subset [n]$ we denote by A_T the matrix restricted to T. In other words, $A_T = (a_{ij})_{i=1,2,\ j\in T}$.

Now we have

Theorem 6.61. *Let G be a simple graph on the vertex set $[n]$, and $S \subset [n]$. Let $A = (a_{ij})_{\substack{i=1,2\\ j=1,\ldots,n}}$ and $B = (b_{ij})_{\substack{i=1,2\\ j=1,\ldots,n}}$ be contingency tables. Then the binomial*

$$f = \prod_{j=1}^{n} x_j^{a_{1j}} y_j^{a_{2j}} - \prod_{j=1}^{n} x_j^{b_{1j}} y_j^{b_{2j}}$$

belongs to $P_S(G) = (\bigcup_{i\in S}\{x_i, y_i\}, J_{\tilde{G}_1}, \ldots, J_{\tilde{G}_{c(S)}})$ if and only if the following condition $(*)$ is satisfied:

$$\sum_{j\in S}(a_{1j} + a_{2j}) \geq 1 \quad \text{and} \quad \sum_{j\in S}(b_{1j} + b_{2j}) \geq 1, \quad \text{or}$$

$\operatorname{supp}(A - B) \subset \bigcup_{i=1}^{c(S)} V(\tilde{G}_i)$, and $A_{V(\tilde{G}_i)}$ and $B_{V(\tilde{G}_i)}$ have equal marginal distribution for $i = 1, \ldots, c(S)$, where $G_1, \ldots, G_{c(S)}$ are the connected components of $G_{[n]\setminus S}$.

Proof. Suppose that $f \in P_S(G)$ and that $\sum_{j\in S}(a_{1j}+a_{2j}) = 0$ or $\sum_{j\in S}(b_{1j}+b_{2j}) = 0$. Say, $\sum_{j\in S}(a_{1j}+a_{2j}) = 0$. Then $\sum_{j\in S}(b_{1j}+b_{2j}) = 0$, too, because otherwise it would follow that

$$\prod_{j=1}^{n} x_j^{a_{1j}} y_j^{a_{2j}} \subset (J_{\tilde{G}_1}, \ldots, J_{\tilde{G}_{c(S)}})R',$$

where R' is the polynomial ring over K in the variables x_j, y_j with $j \notin S$. Since $P' := (J_{\tilde{G}_1}, \ldots, J_{\tilde{G}_{c(S)}})R'$ is a prime ideal, this would imply that one of the variables x_j or y_j with $j \notin S$ would belong to $P_S(G)$, a contradiction.

It follows that $f \in P'$. Let $f = ug$ where u is a monomial and g is a binomial whose monomial terms have greatest common divisor 1. By using the fact that P' is a prime ideal containing no variables we conclude that $g \in P'$. Now we apply Problem 5.8 and deduce that $g \in P''$ with $P'' = (J_{\tilde{G}_1}, \ldots, J_{\tilde{G}_{c(S)}})R''$, where R'' is the polynomial ring over K in the variables x_j, y_j with $j \in \bigcup_{i=1}^{c(S)} V(\tilde{G}_i)$. Thus if $C = (c_{ij})_{\substack{i=1,2 \\ j=1,\ldots,n}}$ and $B = (d_{ij})_{\substack{i=1,2 \\ j=1,\ldots,n}}$ with $g = \prod_{j=1}^n x_j^{c_{1j}} y_j^{c_{2j}} - \prod_{j=1}^n x_j^{d_{1j}} y_j^{d_{2j}}$, then

$$\mathrm{supp}(A - B) = \mathrm{supp}(C - D) \subset \bigcup_{i=1}^{c(S)} V(\tilde{G}_i).$$

As $g \in P''$ it follows from Lemma 6.60 that for $i = 1, \ldots, c(S)$ the tables $C_{V(\tilde{G}_i)}$ and $D_{V(\tilde{G}_i)}$ have equal marginal distributions. Since there exists a matrix E such that $A = C + E$ and $B = D + E$ the same holds true for A and B. Thus condition $(*)$ follows.

Conversely, suppose that $\sum_{j \in S}(a_{1j} + a_{2j}) \geq 1$ and $\sum_{j \in S}(b_{1j} + b_{2j}) \geq 1$. Then $f \in (\bigcup_{j \in S}\{x_j, y_j\})$, and hence $f \in P_S(G)$. On the other hand, if $\mathrm{supp}(A - B) \subset \bigcup_{i=1}^{c(S)} V(\tilde{G}_i)$, and $f = ug$ with u a monomial and g a binomial as in the first part of the proof, then the condition that $A_{V(\tilde{G}_i)}$ and $B_{V(\tilde{G}_i)}$ have equal marginal distribution for $i = 1, \ldots, c(S)$ together with Lemma 6.60 guarantee that $g \in P''$, so that $f \in P_S(G)$. □

Now we are in the position to give numerical conditions for two contingency tables of shape $2 \times n$ to be connected via a given set of moves.

Corollary 6.62. *Let A and B be contingency tables with equal marginal distribution. Furthermore, let $\mathcal{S}$ be a set of moves and I_G the binomial edge ideal whose binomial generators correspond to the moves in $\mathcal{S}$. Then A and B are connected via $\mathcal{S}$, if and only if for each subset $S \subset [n]$ for which each $j \in S$ is a cut-point of $G_{([n]\setminus S)\cup\{j\}}$ the contingency tables A and B satisfy condition $(*)$ of Theorem 6.61.*

Proof. By Theorem 6.53 we have to show that the binomial f which is associated to the pair (A, B) of contingency tables belongs to I_G. Since by Corollary 6.50 the ideal I_G is a radical ideal it follows that $f \in I_G$ if and only if f belongs to all minimal prime ideals of I_G. By Corollary 6.58 these are exactly the prime ideals $P_S(G)$ where each $j \in S$ is a cut-point of $G_{([n]\setminus S)\cup\{j\}}$. What it means that f belongs to such a prime ideal is described in Theorem 6.61. Thus the assertion follows. □

Example 6.63. Let $A = (a_{ij})$ and $B = (b_{ij})$ be two contingency tables of shape 2×4. Let $\mathcal{S}$ be the set of adjacent moves

$$\begin{pmatrix} 1 & -1 & 0 & 0 \\ -1 & 1 & 0 & 0 \end{pmatrix}, \begin{pmatrix} 0 & 1 & -1 & 0 \\ 0 & -1 & 1 & 0 \end{pmatrix}, \begin{pmatrix} 0 & 0 & 1 & -1 \\ 0 & 0 & -1 & 1 \end{pmatrix}.$$

Then we deduce from Example 6.59 and Corollary 6.62 that A and B are connected via $\mathcal{S}$ if and only if the following conditions are satisfied:

(a) $\sum_{j=1}^{4} a_{ij} = \sum_{j=1}^{4} b_{ij}$ for $i = 1, 2$;

(b) $a_{1j} + a_{2j} = b_{1j} + b_{2j}$ for $j = 1, 2, 3, 4$;

(c) either $a_{12} + a_{22} \geq 1$ and $b_{12} + b_{22} \geq 1$, or $a_{ij} = b_{ij}$ for $i, j \leq 2$, and $a_{13} + a_{14} = b_{13} + b_{14}$ and $a_{23} + a_{24} = b_{23} + b_{24}$;

(d) either $a_{13} + a_{23} \geq 1$ and $b_{13} + b_{23} \geq 1$, or $a_{ij} = b_{ij}$ for $i, j \geq 3$, and $a_{11} + a_{12} = b_{11} + b_{12}$ and $a_{21} + a_{22} = b_{21} + b_{22}$.

Problems

Problem 6.1. Let $f_1, \ldots, f_m \in S$ be a regular sequence, and let

$$\epsilon \colon \bigoplus_{i=1}^{m} S e_i \to (f_1, \ldots, f_m)$$

the S-module homomorphism with $\epsilon(e_i) = f_i$ for $i = 1, \ldots, m$. Show that $\operatorname{Ker} \epsilon$ is generated by the elements $f_i e_j - f_j e_i$ with $i < j$.

Problem 6.2. Let R be a hypersurface ring, that is, a ring of the form $R = S/(f)$ where $f \in (x_1, \ldots, x_n)^2$ is a homogeneous polynomial. Show that R is Koszul if and only if f is of degree 2.

Problem 6.3. By using Proposition 6.2 and the result of Problem 6.2 compute the Poincaré series of a hypersurface ring.

Problem 6.4. Prove that a hypersurface ring $R = S/(f)$ where f is a quadratic binomial is strongly Koszul.

Problem 6.5. Show that the 2-squarefree Veronese subalgebra of $S = K[x_1, x_2, x_3, x_4]$ is strongly Koszul.

Problem 6.6. Let $u_1, \ldots, u_m \in S = K[x_1, \ldots, x_n]$ be monomials of same degree d and $R = K[u_1, \ldots, u_m]$ the toric ring endowed with the standard grading given by $R_i = R \cap S_{di}$ for $i \geq 0$. Show that R is strongly Koszul if and only if the ideals $(u_i) \cap (u_j)$ of R are generated in degree 2 for all $i \neq j$.

Problem 6.7. Let R be the 2-squarefree Veronese subalgebra of $S = K[x_1, x_2, x_3, x_4, x_5]$. Show that R is Koszul, but not strongly Koszul.

Problem 6.8. Let $A \subset K[x,y]$ be a monomial subalgebra of $K[x,y]$ containing the set of monomials $\mathcal{M} = \{xy^i\colon\ i \geq 0\}$ but no pure power of y. Show that $\mathcal{M}$ is part of any minimal set of generators of A.

Problem 6.9. Let $B \subset S = K[x_1,\dots,x_n]$ be a sortable set of monomials and $s = (s_1,\dots,s_n)$ an element in $\mathbb{N}^n$. Show that the set

$$B_s = \{x_1^{a_1}\cdots x_n^{a_n} \in B\colon\ a_1 \leq s_1,\dots,a_n \leq s_n\}.$$

is also sortable.

Problem 6.10. Let $L \subset S = K[x_1,\dots,x_n]$ be a set of monomials of degree d. The set L is called an **initial lexsegment** if for all $u \in L$ and monomials $v \in S_d$ with $v >_{\text{lex}} u$ it follows that $v \in L$. Show that initial lexsegments are not always sortable and characterize those which are sortable.

Problem 6.11. By analyzing carefully the arguments in the proof of Theorem 6.17, show that the toric ideal of the squarefree Veronese algebra has a squarefree initial ideal with respect to the reverse lexicographic order.

Problem 6.12. Let P be a finite poset and $\mathcal{I}_r(P)$ the distributive lattice of r-multichains of poset ideals of P, as defined in Subsection 6.3. Let Q_{r-1} be the poset of the integers $1 \leq i \leq r-1$ equipped with the natural order. Then $P \times Q_{r-1}$ is again a poset with the partial order defined componentwise. Show that the distributive lattice $\mathcal{I}_r(P)$ can be naturally identified with the distributive lattice $\mathcal{I}(P \times Q_{r-1})$ of poset ideals of $P \times Q_{r-1}$.

Problem 6.13. A finite poset P is called **pure** if all maximal chains in P have the same length. A well-known theorem of Hibi [**Hi87**] says that the Hibi ring attached to a poset P is Gorenstein if and only if P is pure. Use this fact and Problem 6.12 to deduce that for any given integer $r \geq 2$ one has that $R_2(P)$ is Gorenstein if and only if $R_r(P)$ is Gorenstein.

Problem 6.14. Let K be a field, $S = K[x_1,x_2]$ the polynomial ring over K in two variables with graded maximal ideal $\mathfrak{m} = (x_1,x_2)$. Given an integer $k \geq 1$, compute the presentation ideal of the Rees ring $\mathcal{R}(\mathfrak{m}^k)$.

Problem 6.15. Let K be a field, $S = K[x_1,x_2,\dots,x_n]$ the polynomial ring in n variables over the field K and $I = (x_1^2,x_2^2,\dots,x_n^2) \subset S$. Compute the presentation ideal of the Rees ring $\mathcal{R}(I)$.

Problem 6.16. Let K be a field, $S = K[x_1,x_2,\dots,x_n]$ the polynomial ring in n variables over the field K and $\mathfrak{m} = (x_1,\dots,x_n)$ the graded maximal ideal of S. Show that the presentation ideal of the Rees ring $\mathcal{R}(\mathfrak{m})$ can be interpreted as the ideal $I_2(X)$ of 2-minors of a suitable $2\times n$-matrix. Deduce from this fact that $\mathcal{R}(\mathfrak{m})$ is Cohen–Macaulay. What is the dimension of $\mathcal{R}(\mathfrak{m})$?

Problem 6.17. Let K be a field, $X = (x_{ij})$ an $m \times n$-matrix of indeterminates with $m \leq n$ and $I_m(X) \subset K[X]$ the ideal of maximal minors of X. Consider the ideal J generated by the elements x_{ij} with $i > j$ or $j \geq i + m$ together with the elements $x_{1j} - x_{i,j+i-1}$ with $i = 1, \dots, m$ and $j = 1, \dots, m$. Show that $(K[X]/I_m(X))/J(K[X]/I_m(X)) \cong T/\mathfrak{m}^m$, where $T = K[x_{11}, \dots, x_{1m}]$ and $\mathfrak{m} = (x_{11}, \dots, x_{1m})$ is the graded maximal ideal of T. Conclude that J is generated by a regular sequence of linear forms.

Problem 6.18. Let K be a field, $X = (x_{ij})$ an $m \times n$-matrix of indeterminates and fix a diagonal monomial order $<$ on $K[X]$. Compute the minimal prime ideals of $\mathtt{in}_<(I_t(X))$ in the case that $m = n - 1$ and in the case that $m = 2$ and n is arbitrary. How many minimal prime ideals does one have in each of these cases?

Problem 6.19. Let K be a field, $X = (x_{ij})$ an $m \times n$-matrix of indeterminates, and let A be the K-subalgebra of $K[X]$ generated by the maximal minors of X. Show that A is isomorphic to a polynomial ring if and only if $m = 1$, $m = n$ or $m = n - 1$.

Problem 6.20. Let K be a field, $X = (x_{ij})$ an $m \times n$-matrix of indeterminates and fix a diagonal monomial order $<$ on $K[X]$, and let A be the K-algebra generated by the maximal minors of X. According to Theorem 6.45 the initial algebra $B = \mathtt{in}_<(A)$ is isomorphic to a Hibi ring attached to a suitable poset. In the case that $m = 2$ and $n \geq 2$ is arbitrary, determine this poset and compute the Hilbert series of A.

Problem 6.21. Let K be a field and A the algebra generated by the 2×2-minors of the matrix

$$X = \begin{pmatrix} x_1 & x_2 & x_3 & x_4 & y_1 \\ x_2 & x_3 & x_4 & x_5 & y_2 \end{pmatrix}.$$

Show that the 2×2-minors generating A do not form a Sagbi basis of X with respect to the lexicographic order induced by $x_1 > x_2 > x_3 > x_4 > x_5 > y_1 > y_2$, while the 2×2-minors of the matrix

$$X = \begin{pmatrix} x_1 & x_2 & x_3 & x_4 \\ x_2 & x_3 & x_4 & x_5 \end{pmatrix}$$

form a Sagbi basis of the algebra they generate with respect to the lexicographic order induced by $x_1 > x_2 > x_3 > x_4 > x_5$.

Problem 6.22. Let A be a K-subalgebra of the polynomial ring $S = K[x_1, \dots, x_n]$ and $\{f_1, \dots, f_n\}$ a Sagbi basis of A with respect to a monomial order $<$. Prove that if $\mathtt{in}_<(f_1), \dots, \mathtt{in}_<(f_m)$ are algebraically independent over K, then $f_1, \dots, f_m$ are algebraically independent over K too.

Problem 6.23. Let G be a simple graph on the set $[n]$. With an admissible path

$$\pi : i = i_0, i_1, \ldots, i_r = j$$

where $i < j$, we associate the monomial

$$v_\pi = (\prod_{i_k < i} x_{i_k})(\prod_{i_\ell > j} y_{i_\ell}).$$

Show that the set of binomials

$$\mathcal{G} = \bigcup_{i<j} \{v_\pi f_{ij} : \pi \text{ is an admissible path from } i \text{ to } j\}$$

is a reduced Gröbner basis of J_G with respect to reverse lexicographic order on S where $x_1 > \cdots x_n > y_1 > \cdots > y_n$.

Problem 6.24. Compute the reduced Gröbner basis with respect to (reverse) lexicographic order with $x_1 > \cdots > x_n > y_1 > \cdots > y_n$ for the binomial edge ideal J_G where G is:

(a) the graph with edge set $\{\{1, j\} : 2 \leq j \leq n\}$;
(b) an n-cycle with edge set $\{\{1, 2\}, \{2, 3\}, \ldots, \{n-1, n\}, \{n, 1\}\}$.

Problem 6.25. Let G be the graph displayed in Figure 13. Show that there exists a labeling of its vertices such that the corresponding binomial edge ideal J_G has a quadratic Gröbner basis with respect to the lexicographic order induced by the natural order of the indeterminates.

Figure 13

Problem 6.26. Let G be the graph displayed in Figure 14. Show that for any labeling of its vertices, the corresponding binomial edge ideal J_G does not have a quadratic Gröbner basis with respect to the lexicographic order induced by the natural order of the indeterminates.

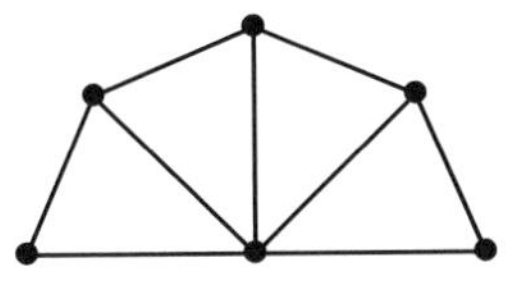

Figure 14

Problem 6.27. Let A be an $m \times n$-matrix of integers with row and column sums equal to zero. Show that A belongs to the sublattice of $\mathbb{Z}^{m\times n}$ which is generated by the matrices of the form $D = (d_{ij})$, where for some integers $1 \leq i_1 < i_2 \leq m$ and $1 \leq j_1 \leq j_2 \leq n$ one has $d_{i_1,j_1} = d_{i_2,j_2} = 1$, $d_{i_1,j_2} = d_{i_2,j_1} = -1$ and $d_{ij} = 0$ otherwise.

Problem 6.28. Consider the matrix

$$X = \begin{pmatrix} x_{11} & x_{12} & x_{13} \\ x_{21} & x_{22} & x_{23} \\ x_{31} & x_{32} & x_{33} \end{pmatrix},$$

and let I be the ideal generated by the following minors:

$$x_{11}x_{22} - x_{12}x_{21},\ x_{12}x_{23} - x_{13}x_{22},\ x_{21}x_{32} - x_{22}x_{31},\ x_{22}x_{33} - x_{23}x_{32}$$

of X. Show that $f = x_{13}x_{21}x_{32} - x_{11}x_{22}x_{33} \notin I$ but $f^2 \in I$. In particular, I is not a radical ideal.

Problem 6.29. Let G be a cycle of length n. Determine the minimal prime ideals of J_G.

Problem 6.30. Describe the conditions for two contingency tables of shape 2×4 to be connected via the moves which correspond to the binomial generators of the binomial edge ideal of a cycle of length 4. How many connected components of contingency tables do we have in this example?

Bibliography

[AL94] W. W. Adams, P. Loustaunau, *An Introduction to Gröbner bases*, Amer. Math. Soc., Providence, RI, 1994.

[BKW93] T. Becker, H. Kredel, V. Weispfenning, *Gröbner bases: a computational approach to commutative algebra*, Springer, 1993.

[BC03] W. Bruns, A. Conca, *Gröbner bases and determinantal ideals*, In: Commutative Algebra, Singularities ansd Computer Algebra, J. Herzog and V. Vuletescu, Eds., NATO Science Series **115**, (2003), 9–66.

[BG09] W. Bruns, J. Gubeladze, *Polytopes, Rings, and K-theory*, Springer Monographs in Mathematics, Springer, 2009.

[BH98] W. Bruns, J. Herzog, *Cohen-Macaulay rings*, Revised Ed., Cambridge University Press, 1998.

[BV88] W. Bruns, U. Vetter, *Determinantal rings*, Lect. Notes Math. **1327**, Springer, 1988.

[Bu65] B. Buchberger, *Ein Algorithmus zum Auffinden der Basiselemente des Restklassenrings*, PhD thesis, Innsbruck, 1965.

[CLO07] D. Cox, J. Little, D. O'Shea, *Ideals, Varieties, and Algorithms: An Introduction to Computational Algebraic Geometry and Commutative Algebra*, Third edition, Springer, 2005.

[CLO05] D. Cox, J. Little, D. O'Shea, *Using Algebraic Geometry*, Second edition, Springer, 2005.

[DES98] P. Diaconis, D. Eisenbud, B. Sturmfels, *Lattice walks and primary decomposition*, Mathematical Essays in Honor of Gian-Carlo Rota, Birkhäuser, Boston, 1998, pp. 173–193.

[E95] D. Eisenbud, *Commutative algebra with a view toward algebraic geometry*, Graduate Texts in Mathematics **150**, Springer, 1995.

[EHV92] D. Eisenbud, C. Huneke, W. Vasconcelos, *Direct methods for primary decomposition*, Invent. Math. **110** (1992), 207–235.

[ES84] D. Eisenbud, B. Sturmfels, *Binomial ideals*, Duke Math. J. **84** (1996), 1–45.

[EHM10] V. Ene, J. Herzog, F. Mohammadi, *Monomial ideals and toric rings of Hibi type arising from a finite poset*, Eur. J. Comb. **32** (2011), 404-421.

[FH11] G. Fløystad, J. Herzog, *Gröbner bases of syzygies and Stanley depth*, J. Algebra **328** (2011), 178–189.

[GTZ88] P. Gianni, B. Trager, G. Zacharias, *Gröbner bases and primary decomposition of polynomials ideals*, J. Symb. Comput. **6** (1988), 149–167.

[G87] G. Glonek, *Some aspects of log linear models*, Thesis, School of Math. Sci., Flinders Univ. of S. Australia, 1987.

[GP02] G.-M. Greuel, G. Pfister, *A Singular Introduction to Commutative Algebra*, Springer, 2002.

[HH05] J. Herzog, T. Hibi, *Distributive lattices, bipartite graphs and Alexander duality*, J. Algebraic Combin. **22** (2005), 289–302.

[HH10] J. Herzog, T. Hibi, *Monomial ideals*, Graduate Texts in Mathematics **260**, Springer, 2010.

[HH11] J. Herzog, T. Hibi, *Ideals generated by 2-adjacent minors*, preprint 2011.

[HHHKR10] J. Herzog, T. Hibi, F. Hreinsdotir, T. Kahle, J. Rauh, *Binomial edge ideals and conditional independence statements*, Adv. Appl. Math. **45** (2010), 317–333.

[HHR00] J. Herzog, T. Hibi, G. Restuccia, *Strongly Koszul algebras*, Math. Scand. **86** (2000), no. 2, 161–178.

[HHV05] J. Herzog, T. Hibi, M. Vladoiu, *Ideals of fiber type and polymatroids*, Osaka J. Math. **42** (2005), 807-829.

[HT96] J. Herzog, N. V. Trung, *Gröbner bases and multiplicity of determinantal and Pfaffian ideals*, Adv. in Math. **96** (1992), 1–37.

[Hi87] T. Hibi, *Distributive lattices, affine semigroup rings and algebras with straightening laws*, Commutative Algebra and Combinatorics (M. Nagata and H. Matsumura, Eds.) Adv. Stud. Pure Math. **11**, North-Holland, Amsterdam, 1987, 93–109.

[Hi92] T. Hibi, *Algebraic Combinatorics on Convex Polytopes*, Carslaw, Glebe, N.S.W., Australia, 1992.

[Ho72] M. Hochster, *Rings of invariants of tori, Cohen-Macaulay rings generated by monomials, and polytopes*, Ann. of Math. **96** (1972), 228–235.

[Ho77] M. Hochster, *Cohen-Macaulay rings, combinatorics, and simplicial complexes*, Proceedings of the Second Oklahoma Ring Theory Conference (March 1976), Marcel-Dekker, New York, 1977, 171-223.

[HS00] S. Hoşten, J. Shapiro, *Primary decomposition of lattice basis ideal*, J.Symbolic Computation **29** (2000), 625–639.

[HS04] S. Hoşten, S. Sullivant, *Ideals of adjacent minors*, J. Algebra **277** (2004), 615–642.

[KM89] D. Kapur, K. Madlener, *A completion procedure for computing a canonical basis for a k-subalgebra*, Computer and Mathematics (Cambridge MA 1989), Springer 1989, 1–11.

[KH06] M. Kokubo, T. Hibi, *Weakly polymatroidal ideals*, Alg. Colloq. **13** (2006), 711-720.

[KR00] M. Kreuzer, L. Robbiano, *Computational Commutative Algebra*, Vol. 1, Springer, 2000.

[KR05] M. Kreuzer, L. Robbiano, *Computational Commutative Algebra*, Vol. 2, Springer, 2005.

[HO99] H. Ohsugi, T. Hibi, *Toric ideals generated by quadratic binomials*, J. Algebra **218** (1999), 509–527.

[Mac01] *Computations in algebraic geometry with Macaulay 2*, D. Eisenbud, D. R. Grayson, M. E. Stillman, B. Sturmfels, Eds., Algorithms and Computations in Mathematics **8**, Springer, 2001.

[M86] H. Matsumura, *Commutative ring theory*, Cambridge University Press, 1986.

[Mi10] E. Miller, *Theory and applications of Lattice Point Methods for Binomial Ideals*, in Combinatorial aspects of Commutative Algebra and Algebraic Geometry, Abel Symposium, 2009.

[MS05] E. Miller, B. Sturmfels, *Combinatorial commutative algebra*, Graduate Texts in Mathematics **227**, Springer, 2005.

[O10] M. Ohtami, *Graphs and ideals generated by some* 2*-minors*, Commun Algebra **39**(3) (2011), 905–917.

[R76] G. A. Reisner, *Cohen-Macaulay quotients of polynomial rings*, Adv. in Math. **21** (1976), 30–49.

[RS90] L. Robbiano, M. Sweedler, *Subalgebra bases*, Commutative Algebra (W. Bruns, A. Simis, Eds.) Lect. Notes Math. **1430**, Springer, 1990, 61–87.

[Sc80] F.-O. Schreyer, *Die Berechnung von Syzygien mit dem verallgemeinerten Weierstrass'schen Divisionssatz*, Diploma Thesis, University of Hamburg, Germany.

[S75] R. Stanley, *The upper bound conjecture and Cohen-Macaulay rings*, Stud. Appl. Math. **54** (1975), 135–142.

[S96] R. Stanley, *Combinatorics and commutative algebra*, Second edition, Progress in Mathematics, **41**, Birkhuser Boston, Inc., Boston, MA, 1996.

[St90] B. Sturmfels, *Gröbner bases and Stanley decompositions of determinantal rings*, Math. Z. **205** (1990), 137–144.

[St95] B. Sturmfels, *Gröbner Bases and Convex Polytopes*, Amer. Math. Soc., Providence, RI, 1995.

[V90] R. Villarreal, *Cohen-Macaulay graphs*, Manuscripta Math. **66** (1990), 277–293.

[V01] R.H. Villarreal, *Monomial algebras*, Marcel Dekker, 2001.

Index